Nothing Left in My Hands

Nothing Left in My Hands

THE ISSEI OF A RURAL CALIFORNIA TOWN, 1900–1942

Kazuko Nakane

Foreword by Naomi Hirahara

Heyday Books, Berkeley, California
BayTree Books

This book was made possible in part by a generous grant from the BayTree Fund.

Originally published by Young Pine Press in 1985

Library of Congress Cataloging-in-Publication Data
Nakane, Kazuko.
Nothing left in my hands : the Issei of a rural California town, 1900-1942 / Kazuko Nakane ; foreword by Naomi Hirahara.
p. cm. -- (BayTree books)
Originally published: Seattle, Wash. :Young Pine Press, c1985.
Includes bibliographical references.
ISBN 978-1-59714-109-3 (pbk. : alk. paper)
1. Japanese Americans--California--Pajaro River Valley--History--20th century. 2. Japanese Americans--California--Watsonville--History--20th century. 3. Japanese Americans--California--Watsonville--Interviews. 4. Immigrants--California--Watsonville--Interviews. 5. Farmers--California--Watsonville--Interviews. 6. Community life--California--Watsonville--History--20th century. 7. Agriculture--California--Pajaro River Valley--History--20th century. 8. Agriculture--California--Watsonville--History--20th century. 9. Watsonville (Calif.)--History--20th century. 10. Pajaro River Valley (Calif.)--History--20th century. I. Title.
F868.S3N35 2009
979.4'71004956--dc22

2008023793

Heyday Books is committed to preserving ancient forests and natural resources. We elected to print this title on 30% post consumer recycled paper, processed chlorine free. As a result, for this printing, we have saved:

2 Trees (40' tall and 6-8" diameter)
797 Gallons of Wastewater
2 million BTU's of Total Energy
102 Pounds of Solid Waste
192 Pounds of Greenhouse Gases

Heyday Books made this paper choice because our printer, Thomson-Shore, Inc., is a member of Green Press Initiative, a nonprofit program dedicated to supporting authors, publishers, and suppliers in their efforts to reduce their use of fiber obtained from endangered forests.

For more information, visit www.greenpressinitiative.org

Environmental impact estimates were made using the Environmental Defense Paper Calculator. For more information visit: www.papercalculator.org.

Cover Design by Rebecca LeGates
Cover Photo by Dorothea Lange, courtesy of The Bancroft Library, University of California, Berkeley, War Relocation Authority Photographs, Series 14, Volume 61, Section G, WRA no. C-315
Interior Design by Leigh McLellan
Maps drafted by Ray Fujii
Printing and Binding by Thomson-Shore, Dexter, MI

Orders, inquiries, and correspondence should be addressed to:
Heyday Books,
P. O. Box 9145, Berkeley, CA 94709
(510) 549-3564, Fax (510) 549-1889
www.heydaybooks.com

10 9 8 7 6 5 4 3 2 1

Contents

Foreword

By Naomi Hirahara

Although my parents and I visited Watsonville every summer during my childhood, I only became familiar with the name—Pajaro Valley—of this region on the central California coast much later, in my thirties. Before then, Watsonville was just the *inaka,* the country, where we would travel north several hours from Los Angeles in my father's white van that carried gardener's tools most of the year. My father was born in Watsonville but had moved to Japan after his grandfather was killed by a horse on the farm. After World War II, my father returned to Watsonville, in his late teens, and before making his way to his new life in Southern California lived with his aunt and uncle and many cousins in a pristine white Victorian house off of Highway 1.

For this city girl, Watsonville was a magical place. The rows of lettuce around my relatives' house resembled the heads of giant groundhogs surveying the rich soil. Since it was summer and in the sixties, before the off-season varieties had been fully developed, strawberries were only found within the confines of plastic containers in the large freezer in the shed. My distant cousins and I ran around inside and outside the large house. One favorite spot for all the grandchildren, great-nephews, and great-nieces was just on the side of the curved staircase on the first floor, where comic books were stacked like a pile of gold. Just where my great-aunt got all those comic books—and they were recent ones, too—was unknown. They seemed to just secretly appear in this country farmhouse.

As I grew older, I began to notice the name "Watsonville, California" affixed to plastic strawberry containers, or shells, but didn't think much of it. Watsonville, Castroville, Salinas, Gilroy—they seemed like the generic *inaka* where fruits and vegetables came from, like a giant icebox serving the city folk. Then, in college in Northern California, I caught the liberation theology fever and went on an overnighter with a group led by our campus pastor to a farm in—where else?—Watsonville. We slept in the home of a farm worker, visited the local United Farm Workers (UFW) field office, and walked amidst mushroom and cauliflower fields. This was a different Watsonville than I had ever known. I didn't attempt to reconcile the childhood memories of sweet frozen strawberries in the packinghouse freezer with the more political and complex issues of labor in a farm community. In my mind, the two worlds—family farm and migrant farm workers—were parallel, unable to coexist separately, like two different crops on the same piece of land.

Almost two decades later, I visit Watsonville again, but this time as a researcher for a prominent Nisei farmer's memoir. My relatives' beautiful Victorian home has been abandoned, heavily damaged by years of rain. The devastating Loma Prieta earthquake in 1989 leveled downtown buildings, and ten year later, gallant yet scattershot efforts at reconstruction are evident on Main Street. I found that to do research in the area, I couldn't stop in at one historic society or one religious organization. People were not only separated by class, race, and cemeteries, but by religion and family connections. Some old-timers were still around to tell me how the seasonal rains would flood their homes and businesses, causing families to move merchandise and even pianos onto the second floor. I heard accounts of migrant farmers moving from one acreage to another, creating patchwork strawberry farms in light of alien land laws outlawing the Japanese from buying land. Yet I needed something more definitive about the Japanese American presence in the area.

There were individual oral histories filed in local and college libraries and Sandy Lydon's *The Japanese in the Monterey Bay Region,* which provided an important timeline of events and photos before and after World War II. But beyond that, my search did not result in much.

I did discover one invaluable source back in a library in Los Angeles—a copy of *Nothing Left in My Hands,* published by a small publisher, Young Pine Press. It was a slim volume, barely over a hundred pages, yet carefully constructed, as if each word had been selected from a handset type box. I was instantly captivated by its simplicity, earnestness, and deep research. The only problem was its availability. Out of print, it seemed even unavailable for sale as a used book. The library where I found it only had it as a noncirculating reference book.

So I took notes, photocopied excerpts, and soaked in its elegant and simple tone while sitting at a desk in the library. *Nothing Left in My Hands,* I thought, would be just one of those precious stones in a few libraries, waiting to be excavated. And I personally was glad to have experienced it.

Almost ten years later, an email comes. Heyday Books is asking me to fashion a foreword for the reprinting of this book, *Nothing Left in My Hands.* I'm delighted. It is no surprise that Heyday Books, with its commitment to California and rural history, would be the one to give a second life to Kazuko Nakane's book and her material.

The value of *Nothing Left in My Hands* is multiple. First of all, it's one of the few published sources of pre–World War II Issei historic accounts of the Pajaro Valley and farming communities. At her foundation are many familiar sources, in both English and Japanese. Being a native from Japan and bilingual, Nakane is able to deftly weave and condense the important facts to provide a good portrait of the region from the early 1900s to 1940.

Appropriately peppering her narrative are excerpts from oral history interviews she conducted with predominantly Japanese-speaking Issei. While some scholars are suspicious of including these types of accounts in history books, I've always acknowledged the power of such information collection, especially when the interviews involve immigrants and are conducted in-language. Multiple factors, notwithstanding the incarceration of Japanese Americans during World War II, led those of Japanese ancestry to distance themselves from the first tongue of the Issei, the Japanese language. As a result, a gulf has emerged between the first and third generations, causing the latter to depend only on secondhand recollections to piece together what really happened before the war.

These interviews are not comprehensive, yet they are wide enough to provide those essential concrete details that give the text life. It is obvious that Nakane respects her subjects immensely. In keeping with her careful work, *Nothing Left in My Hands* is not purely a celebration, but a documentation of harsh realities. Instances of gambling, prostitution, and labor protests are all meticulously documented.

Geography and the changing continuum of agriculture are also important. Through the study of Pajaro Valley, you can see the transfer of labor and immigration patterns. The contributions of the early Chinese and Dalmatians (Croatians from Yugoslavia) and later Pilipino and Mexican laborers serve as bookends to the entry of agricultural workers from Japan. The Japanese labor force has steadily transitioned from *buranketto katsuji,* migrants carrying blankets on their shoulders, to sharecroppers and cash tenant farmers and to the independent farmers of today.

Women and mothers are not forgotten. Again, through the practice of oral history, Nakane has also provided the reader with searing snapshots of the lives of Issei women who traveled across the Pacific to marry men who were virtually strangers.

Certainly other agricultural areas along the Pacific Coast share similar stories. But there's something about Watsonville—perhaps its proximity to the ocean or its gentle wildness—that invites comparisons to the coast of Japan. Or perhaps it's my own childhood nostalgia that only deepens my affinity for the area.

While many outlying communities closer to or within the Silicon Valley have fallen to high-tech development, Watsonville has held onto its agricultural roots—albeit with the pressure of large housing projects pressing against its boundaries. Because of this long tradition, the tale of Pajaro Valley has added weight. Indeed, as we read the accounts of these pre–World War II farmers, other faces—the tired, calloused men and women escaping the Dust Bowl in the thirties, the stooped images of today's strawberry pickers in the fields—seem to have more definition. *Nothing Left in My Hands* is the story of one specific ethnic group during a window of time, yet a story that explains much of what came before and what came after.

Preface to the New Edition

The Fruits of Their Labors

It was on December 19, 1974, that I landed at the Oakland Airport, first arriving in America. I did not know what to expect for a new life here with my husband, Alan Lau, whom I had met in Kyoto City, Japan. The apartment we got in Capitola remained spare and abundantly spacious, without furniture for a long time. I bicycled to Cabrillo College until I transferred to the University of California, Santa Cruz. I also bicycled to my first job, as a dishwasher at a convalescent hospital, arriving before four o'clock in the morning on weekends. The young women who worked there wore white shoes and were chain smokers, and they drank many cups of coffee to help them make it through each day. Capitola was a quaint tourist town, a place for retirees. On the rocky beach below the cliff, referred to as China Beach, Chinese immigrants had had a fishing village until the anti-Chinese sentiment of the 1880s chased them away. By the time I arrived, it was a quiet beach, washed away by years of waves from the Pacific Ocean.

Here in the United States, I was at the bottom of the social strata, called "fresh off the boat," and I knew I had to struggle to make this place my home. This was quite different from my previous stay in Germany, where I was treated more like an exotic and precious guest from the Far East.

But the 1970s were also a time of surging ethnic pride. At Cabrillo College, the first class I took was Asian American history taught by Sandy Lydon, who is now a noted local historian. One of my first friends, the late Sharon Lew, spoke to everybody to organize an Asian American group on campus. She arranged for us to take part in the first pilgrimage to Tule Lake, one of the camps where Japanese Americans were incarcerated during World War II. A Nisei and former Tule Lake internee who joined this pilgrimage shared with us his story, to make that history a vivid personal experience. Then, when I was a student at UC Santa Cruz, I participated in an oral history project led by the late Pat Sumi, meeting Issei and older Nisei in Watsonville. They possessed an amazing spirit as survivors, and like me they struggled to make this place their home.

Watsonville is located just south of Santa Cruz, ninety-five miles from San Francisco. Now Main Street is filled with Latinos, largely from Mexico, who make up three-fourths of the city's total population of forty-four thousand. One can hardly imagine that once, at the far end of Main Street and near the bridge where the river flooded many times, stood the residences of the many Issei who lived there before WWII, when they were forced to leave behind their homes and spend the duration of the war in internment camps miles away. The 2000 census still records small numbers of Asian Americans living in this region.

Chinese Americans, Japanese Americans, and Filipino Americans all made telling contributions to the development of agriculture in Watsonville. Young men from China came to California during the Gold Rush, in the 1850s, to farm sugar beets and later work on strawberry farms. Young men from the Philippines came to the area, starting in the late 1920s, to harvest lettuce. The name of Watsonville was carved onto the pages of United States history with the 1930 Watsonville Riots, one of the worst anti-Filipino movements in California history. In January of that year, Filipino lettuce pickers were attacked at a taxi-dance hall and at their homes for five terrifying days. One person was killed. (In spite of this, Filipino workers succeeded in organizing a union in Salinas that same year.) Without these immigrants' struggles and their desire to make it in this country, Watsonville would not be known for its apple cider, strawberries, raspberries, and lettuce. Their hard labor harvested the crops and assured the prosperity of California agriculture, providing a valuable source of food to the rest of America for many years.

The Japanese American experience in the area began with Issei who came from Japan and settled here. There are many publications that document their hardship and investigate their history, particularly during the internment. My hope for this book is to give Issei and older Nisei their rightful place as pioneers who laid the groundwork for subsequent generations of immigrants. Farmwork was one of a few options available to Issei: at one time over 50 percent of bachelors took demanding and unpredictable work in agriculture, hoping to someday gain the financial leverage to succeed as independent landowners. These young men from Japan were familiar with labor-intensive, small-scale farming, and they first worked as seasonal laborers in the area, trying very hard to rent small patches of land to grow their own crops. They hoped to one day have families and eventually own the land. It was difficult but they struggled to make this place their home.

It has been over twenty years since I interviewed Issei and older Nisei in Watsonville. Most of them are long gone and only live on in memories and through this book. Of all the informants I interviewed, there are only two still living, Mrs. Hiroko Shikuma and Mrs. Kazue Murata, both widows, who worked alongside their husbands and are now well into their nineties.

Now there are more regional agricultural studies of Japanese American communities, but this was one of the earliest to give voice to Issei, both men and women, who remembered life at the turn of the century. It is my hope that this book will convey their voices to the next generation of Japanese Americans. I also want their stories to help others understand the complexity of United States history. Each succeeding immigrant generation and their children have brought diversity, new energy, and vibrancy to American culture, and helped make this country unique.

Kazuko Nakane

Preface to the First Edition

It has been over five years since I originally began this project, and I am happy to see this book being published. Now I realize how much I have learned and have been nurtured by this experience.

The Asian American Oral History Project started in 1978 as a class project at the University of California at Santa Cruz. In the early stages I worked together with Kathy Hattori interviewing the Issei of the Pajaro Valley, California. Unfortunately, the class ended without the project being finished. Because I was the translator and one of the interviewers, I decided to continue it by myself. The project became known as the Issei Oral History Project in Watsonville, California, and deals only with the early history of Japanese Americans in that area.

I would like to thank everybody who helped me to make this book a reality. My special thanks to the following people who one way or another advised me or supported me on this project: Y. W. Abiko, librarian Tom Bolling, Naomi Brown, Prof. Maclyn Burg, Carol Champion, Ron Chew, Sue Chin, Ray Fujii, Lori Fukuda, Chris Huie, Prof. Yuji Ichioka, Kathy Torigoe Ichinaga, Prof. Sue-Ellen Jacobs, Rev. Sumio Koga, Karen and Ernie Kuwahara, Charles Leong, Hoy and Margaret Lew, Elaine Murakami, Mona Nagai, Seizo Oka, the Proctors, Harumi Sakurada, Karen Seriguchi, Hiroko and Mack Shikuma, Mario Singleterry, Peter Stolich, Rev. Heihachiro Takarabe, Shirley and Jeff Tagami, Hiroyuki Tani, Asako Tokuno,

Mayumi Tsutakawa, and oral historian Karyl Winn. Thanks also to Alan Lau for his encouragement.

I benefited greatly from the Shiryōshitsu (Archive of Japanese American History at California First Bank), Shin Nichibei Shimbunsha in San Francisco, and Media Services, the map library, and Special Collections at the University of California at Santa Cruz. I would also like to mention that, I was able to recreate a map of Watsonville in 1920 with the help of Frank Enomoto, Charles Iwami, Harry Yagi, and several Issei whom I interviewed. Finally, I would like to thank all the Issei and Nisei in Watsonville whose history and lives I have documented here. I have tried my best to tell their history as accurately as possible but in the end I assume full responsibility for any omission or mistakes.

Kazuko Nakane

Introduction

This book is about the Issei in the Pajaro Valley, a rural community in central California. The Issei are early immigrants from Japan who are now old or have already disappeared into the past.

In 1880 the census reported 148 Japanese living in the United States. Emigration from Japan became legal in 1885, but due to the Immigration Exclusion Act in 1924, Japanese emigration to the United States lasted less than half a century. In 1952, the Walter-McCarran Immigration and Naturalization Act allowed new immigrants from Japan into the U.S., extending a token immigration quota to Asian nations. In 1965, immigration bills finally eliminated race, creed, and nationality as a basis for excluding immigrants. I will not discuss in this book the new immigrants since 1952 or the "Shin Issei" (New Issei) because their experience in the U.S. differs greatly from that of the early Issei.

Under the promise of the Gentlemen's Agreement with the U.S. in 1907, the Japanese government stopped issuing visas, especially to laborers wanting to come to the U.S. The Agreement allowed wives and children of immigrants already in the U.S. to emigrate. Yobiyose are the children of the pioneer Issei, called by their parents to help with work in the U.S.

Many Yobiyose were in their early teens when they landed at the ports of the new land. They are now old men and women and have become the majority of the surviving Issei. Most pioneer Issei, who broke new ground with their vigorous spirit in this young land, are gone. Some pioneer Issei survive like withered trees washed clean by the waves of hardship that followed one after another. It is not too late to learn about them. We can still ask these Issei about their lives, though their numbers are small.

Because of the language barrier, the experiences of the Issei have not been adequately studied. Moreover, documentation on the Issei in rural communities is rare since studies about them usually center on the large cities, such as Los Angeles and Seattle. As a result, lives of the rural Issei are not well understood. Assumptions tend to take the place of real understanding. Very few people have considered and treated the Issei as equal members of American society. But they are a part of U.S. history, and we should understand their lives and appreciate their contributions to this country.

It may be easy to think of the Issei as outsiders since they kept the language, general cultural values and heritage of the country they were born in. Until World War II, the Issei were forced to live in their own communities by constantly issued restrictive laws, hard working conditions that often required mutual dependence, and discrimination against which they had to defend themselves. Also, Issei brides, who seemed to outnumber other Issei immigrants after 1907, came straight to the Japanese rural community from the port and had few opportunities to interact with white people in English. Most importantly, until 1952, the law prohibited the Issei from becoming naturalized citizens of the U.S.

Unless they came to know them personally, the majority of people in the United States assume that the Issei, who strongly retain their Japanese heritage, are foreigners, simply persons from Japan. Japanese from Japan tend to regard the Issei as a part of Japan. They consider the experience of the Issei as Japanese emigrant history, rather than Japanese American history.

Recently, Japanese Americans, particularly the young generation, have expressed a concern about discovering their "roots." They express their interest in the Issei and respect them as forebearers. But some Japanese Americans, who are citizens of the United States by birth, still regard the Issei as part of Japan. The Issei represent Japan to those who, in many cases, have never seen that country. Kibei Nisei, citizens of the United States by birth but educated in Japan, also face this complexity of their status.

Whether the Issei consider themselves a part of Japan of a part of the United States is also complicated. The decision could be a purely personal choice, but one cannot disregard the influence of the hardships and discrimination they suffered throughout their lives here. Presently, they have either citizenship or permanent resident rights in the U.S. Nevertheless, the Issei still distinguish themselves from white Americans, calling the latter "Hakujin" (white) or "America-jin" (American people). I once heard an Issei use the term "gaijin" (foreigner) to describe the white Americans too. Who, then, are the Issei?

Early Agriculture in Pajaro Valley

Pajaro Valley

California agriculture could not have developed without the immigrants who worked in the fields. They stooped down and squatted on their haunches over the earth, their dark hands harvesting the crops. The Pajaro Valley is one of the richest agricultural areas in California, producing potatoes, apples, berries, lettuce, and other crops.

The two main cities in Santa Cruz County, Santa Cruz and Watsonville, are quite different. The former is largely residential with a predominantly white population. It is a university town as well as a summer resort. The latter is an agricultural town and residence for a multi-ethnic population. According to the 1980 census, in the Watsonville division, the white population was 33,761 (which includes those of Spanish origin, 19,059); Asian and Pacific Islander, 2,549; American Indian, Eskimo, Aleut, 457; Black, 223; and others, 11,107. The population totaled 48,097.

Once you step into the town of Watsonville, the Hispanic population appears to outnumber the other groups. Many Mexicans move into the Valley as seasonal farm laborers from spring to fall. Every Sunday, after attending church, families of Hispanics dress up and crowd Main Street in downtown Watsonville. The mingling sounds of Spanish and English give life to the street. From the small park at the center of the city, down to the east where the bridge crosses over to Monterey County, many bars and restaurants line Main Street. Turning off north close to the bridge, old porched houses still stand. One block north on Union Street, parallel to Main Street, there is a small Japanese store, indicative of the inconspicuous Japanese American population scattered throughout the Pajaro Valley.

This store is just like any general store in the Japanese countryside, where you can find many kinds of goods. If you ask for a box of incense sticks, a Nisei woman will climb up on a wooden stool to reach the top shelf and dust it off before handing it to you. Fresh fish and tofu are available. Packages of rice crackers, potato chips, apple pies, bottles of soy sauce, and sake are stacked up leaving space for customers to walk by.

Of course, Japanese is spoken. But this is the Pajaro Valley where Tagalog, Visayan, Spanish, and English are also spoken. Nisei housewives greet friends, asking, "How have you been?" An old bachelor with dark skin buys several packages of cigarettes and a few other items. Without asking his name, the grocer just fills in a credit account sheet, pulling it out from under the counter.

Before World War II, many Pilipinos worked in the fields, bending down among the rows of lettuce, the sun covering their black shadows. Before the Pilipinos, the Issei were the immigrant laborers in the fields, who touched and nurtured the crops with their hands. It is no exaggeration to say that California agriculture cannot be discussed without mentioning the contributions of immigrants, including the Issei. The contribution of the Issei in the Pajaro Valley was certainly significant in the Valley's fruit industry. At first, the Issei provided their strength as seasonal laborers, only slowly moving their way up as sharecroppers, leased farmers, and occasional landowners. All this was accomplished as they struggled against hostile elements.

Early Agriculture

The Pajaro Valley is located in Monterey Bay, over two hours drive south of San Francisco. The Valley covers about 50,000 acres and lies on the border between Monterey and Santa Cruz counties. The largest part, including the town of Watsonville, is in Santa Cruz

County, but the east side of the Valley, about 15 miles long and from 6 to 8 miles wide, is in Monterey County. The Valley has long been one of the treasure houses of California agriculture. The neighboring Salinas Valley in Monterey County, well known for lettuce, is large. Pajaro Valley, on the contrary, is small and intimate, defined by the coastal range of mountains on the east and the white waves of Monterey Bay on the west. The land is fertile, and the climate is moderate year round without marked extremes of heat and cold. The history of agriculture in the rich Valley started long before 1892 when the Japanese arrived.

After the Gold Rush, the population of California increased dramatically, which brought sudden changes to Pajaro Valley. In 1850, the U.S. Census reported a population of 643 in Santa Cruz County, mostly ranch owners and their Indian workers. In 1860, the population jumped to 4,944. According to Edward Martin (1964, p. 9), the population of the Pajaro Valley could not have exceeded 50 in 1851. It did not take long for a carefree, pastoral land to become farmland. Homes with gardens and orchards appeared everywhere. When potatoes became a successful crop and the whole Valley became a vast potato field in 1853, large numbers of squatters moved in. Though the potato boom was shortlived, many farmers insisted on raising potatoes as well as wheat and other grains. The population in Santa Cruz County swelled to 8,743 in 1870 and 12,802 in 1880, according to the census.

Fruit Culture

Among the crops that made the name of Watsonville famous in Pajaro Valley, fruits were very important, especially apples and strawberries. Many Issei worked for apple-packing houses and in strawberry cultivation in Watsonville. However, the Japanese were not pioneers in these fields.

According to Carey McWilliams (1939, p. 59), the 1870s in California agriculture were the years of "Fruit Versus Wheat." After the 1870s, although California was predominantly producing grains, fruit slowly started taking the place of wheat and other grain crops in many areas. The Pajaro Valley was one of the areas where this change occurred.

Fruit cultivation could not progress without a germination

period. Though the numbers were small, the growing of apples and the planting of strawberries started early in the Valley. Besides those fruit trees originally growing in backyards, Fred W. Atkinson, in *100 Years in the Pajaro Valley from 1769 to 1868* (1934, pp. 67-68), said that the first commercial fruits, mostly apples, were planted in 13 acres of orchards by Isaac Williams and in another area by Judge R. F. Peckham in 1858. He adds that, "James Waters, who came to Santa Cruz County in 1855 and the Pajaro Valley in 1859, put out 1900 trees on the William Birlem place" in 1861. Martin (1911, p. 109) reported less than 50 acres of fruit trees, and Atkinson (op. cit., p. 67), about 60 acres of total planting in 1860. Atkinson also noted, "As late as 1864 any rancher who planted more than a few trees was ridiculed by his neighbors; one family, they said, couldn't use so much fruit." This rural area was incorporated into the town of Watsonville by the legislature in 1867.

Apples from the Santa Cruz mountains were known for their superior quality, and Martinelli's apple soda industry was established in Watsonville in 1868 (*Watsonville Register-Pajaronian*, Sept. 26, 1979). In 1866, the number of apple trees in Santa Cruz County, 41,479, was far fewer than that of other counties: Santa Clara County had 210,000, Alameda County had 170,473, and Shasta County, 231,251 (California State Agricultural Society, 1868, p. 559). Apples would not yet bring fame to Santa Cruz County.

The history of the Issei in Watsonville and the history of strawberry cultivation are inseparable, but strawberries were part of the Pajaro Valley even before the arrival of the Issei. *A History of the Strawberry* by Stephen Wilhelm and James E. Sagen (1974) covers strawberry history in the Valley in quite a bit of detail. A kind of strawberry indigenous to the Monterey Bay area was enjoyed by the native Indian people who would picnic for a week or more to feast on strawberries. As settlers moved in and orchards flourished, the strawberry was also planted. Swain Ranch, in the neighboring town of Santa Cruz, was noted for its large strawberries in 1859. Though small-scale, the first commercial strawberry farm in the Valley was the Gilkey farm in the Vega district of Monterey County. The farm was established in 1865. Others followed a few years later. Commercial strawberries were sold to the local market, often with difficulty.

Berry cultivation in Santa Cruz County in 1866 was still small-scale (California State Agricultural Society, op. cit., pp. 560-561). The number of gooseberry bushes was 1,332; raspberry bushes, 3,088; and strawberry vines, 67,450. Other nearby counties in California at the same time had much more intensive cultivation, for example, 12,000,000 vines in Santa Clara County and 6,887,430 in Alameda County.

To produce fruit commercially, irrigation is necessary, especially for strawberries. According to Wilhelm and Sagen (op. cit., p. 175), Watsonville grower-partners employed windmills to pump water for irrigation in 1875. Four years later, the Watsonville Water Works installed flumes from its Corralitos reservoir to dispose of overflow water. High, wooden flumes were also set up at Lake Farm, channeling water from Laguna Grande, a large lake (Harrison, op. cit., p. 30). Around 1890, it was possible to raise 100 acres of strawberries, 40 acres of raspberries, and a few acres of blackberries in Lake Farm itself. This regular supply of water provided the basis for industrial strawberry cultivation.

San Francisco was by far the largest city nearby, increasing its population to 56,802 in 1860; 149,473 in 1870; and 223,959 in 1880. The biggest town in Pajaro Valley was Watsonville, which had a population of approximately 2,000 in 1871, according to Edward Martin (1964, p. 8). Until the railroad system became established between Pajaro Valley and San Francisco, the transportation of commercial crops was hindered. In the early years of Pajaro Valley, boats were used as a means of shipping crops. "Freight was conveyed to the vessels by means of surf boats," on a landing near the beach, though later a substantial wharf was built where the vessels could load directly (ibid., p. 15).

The first train arrived at Pajaro, an area across the Pajaro River from the town of Watsonville, on Nov. 27, 1871, according to the *Watsonville Pajaronian* (Nov. 30, 1871).

> At length the long expected, much talked of Pajaro Branch of the Southern Pacific Railroad has reached its terminus. For nearly three months we have been informed from various sources, that this desirable event would take place in "ten days." Consistent with this expectation we have made business arrangements, and have been disappointed, and some have been accommodated by the "construction train" and so relieved in their disappointment.

The beginning issues of the local paper, the *Watsonville Pajaronian* (which printed its first issue on March 5, 1868), talked about expectations raised by the railroad. People in the Pajaro Valley had waited a long time. According to the Nov. 30, 1871 issue of the *Watsonville Pajaronian*, the train left San Francisco at 8:10 a.m., arrived at Pajaro at 1:30 p.m., and returning, left Pajaro at noon and arrived at San Francisco at 5:30 p.m. Because of the train, businesses could increase their production, knowing they had a big market for their crops. Toward the end of the 1870s, the "Watsonville-to-San Francisco strawberry-shipping boom" started. Cultivation of strawberries slowly increased in the Pajaro Valley: 42 acres in 1881, 45 in 1882, 118 in 1883, 185 in 1884, and 268 in 1885 (Wilhelm and Sagen, op. cit., p. 176).

Another Advancement in Fruit Culture

The development of fruit agriculture in California owes much to a succession of immigrant groups who provided the labor force to work the fields. Landowners, unable to harvest the large stretches of land by themselves, relied on laborers to do the actual farming. In succession, groups of Chinese, Japanese, Pilipino, and Mexican laborers were employed, along with many others. Contractors held an important position, serving as interpreters and mediators between laborers and employers. Within a few years after the end of Mexican rule in California, evidence of a labor contracting system began to appear in agriculture, according to Lloyd H. Fisher (1953, p. 20). Significantly, the system has lasted to this day.

Fisher stresses the relationship between harvesting and seasonal labor in his book. Before World War II, it was inconceivable that machines could be used to harvest fruit crops, unlike with grains. Therefore, the harvesting of fruit required the employment of large numbers of laborers for short, limited periods of time. The same was true for packing. Fresh fruits must be picked and sold as soon as possible because the quality of the crops falls as time passes. Immigrants were contracted as seasonal workers. Since many of these immigrants – Chinese, Japanese, and Pilipino – were single, their lifestyle was flexible and they could move from crop to crop whenever and wherever they were needed. When they settled down in an area permanently, new groups of immigrants took their place as migratory laborers.

The size of farms in Santa Cruz County was small and remained small, in contrast to agriculture in other places in California. According to the 1880 census, the average farm in Santa Cruz County was 189 acres, compared to the average farm in California of 462 acres. As late as 1940, the average farm in Santa Cruz County was 61.4 acres, compared to 230.1 acres for the average farm in the state.

In general, the local farmers owned their land in the small Pajaro Valley. Around 1890, E. S. Harrison said, "Small landholdings are the rule here, and diversified farming follows as a natural result" (op. cit., p. 25). There was a chance that Pajaro Valley might have come under single management and turned to the production of a single crop after the Spreckels sugar factory was constructed in the valley in 1889. But the factory folded and moved to Salinas after about ten years of production. The trend toward smaller plots of land, owned and managed by local farmers, continued.

Appearance of the Issei

The appearance of two immigrant groups in the Pajaro Valley, the Japanese and the Dalmations, coincided with the new developments in fruit agriculture. These new immigrants were hard workers and skilled laborers, as were the Chinese who had come to the Valley at the end of the 1860s and were an important labor force. By 1900, the switch to fruit cultivation became obvious. Apple cultivation spread through the Valley first. The *San Francisco Chronicle* mentioned the growing numbers of apples in Watsonville around 1900 in reports from the different counties:

> A wonderful progress has been made in the apple industry, especially in the Pajaro Valley, where 8,000 acres are devoted to orchards, in which are 500,000 apple trees, the great majority of which were planted since 1890 (Dec. 30, 1900, p. 23).

There was no mention of strawberry cultivation in the article, but *Pacific Rural Press* (Jan. 19, 1901) could not ignore it: "There is going to be a largely increased acreage in strawberries in the Pajaro Valley this year." The increase in the number of orchard trees, especially apple, coincided with strawberries in the early days as strawberries were commonly planted around trees in young orchards. Though there were few new orchards in later years, peas and other vegetables were occasionally planted until the young trees grew productive. But strawberry cultivation was not recommended because some said the excessive irrigation would ruin the proper growth of the tree roots.

On Aug. 9, 1902, the *San Francisco Chronicle* mentioned:

> Although apples lead, and although there has been a great planting in this fruit during the past ten years, berries have, all things considered, the prominent place as a profitable crop. The yield of strawberries is enormous. It will startle the Eastern farmer to hear that the growers pick these berries during nearly ten months of the year. (p. 10).

The names of Dalmations (people from the province of Croatia in Yugoslavia) and Japanese were not made known here, but Reports of the Immigration Commission (1911b, p. 442) indicated their contribution to the cultivation of berries and apples:

> The Japanese and Dalmations have assisted in producing the changes introduced in the kinds of crops grown. The former, being unusually skillful berry growers, have had something to do with the expansion of the production of berries until much of the land is thus employed, whereas before their influx, little of it was so used. The latter have done much to encourage the growing of apples.

Agricultural reports from 1880 to 1940 reveal that Santa Cruz County faced a dramatic decline in grain production near the turn of the century. Production of barley remained rather steady, but that of oats dwindled by almost one-third from 1900 to 1910. Wheat production in 1900 was only one-seventh of what it had been in 1890.

On the other hand, there was a great increase in fruit production. The number of apricot, cherry, and other fruit trees increased. Most of all, the number of apple trees increased from 109,828 in 1890 to 557,361 in 1900. The strawberry acreage in Santa Cruz County in 1900 was 520, producing 2,989,100 quarts. The total value of small fruits in Santa Cruz County was $164,953; surprisingly, it was the highest of all counties in California.

According to the 1910 agricultural census report, in Santa Cruz County the total value of agricultural production was $2,408,435; the value of cereals, other grains and seeds was $118,874 (5 percent); hay and forage, $337,572 (14 percent); vegetables, $164,518 (7 percent); and fruits and nuts, $1,656,212 (69 percent of the total value). The agricultural development in Pajaro Valley brought about by the Dalmations and Japanese, in the footsteps of the Chinese, was enormous.

Leaving the Village

The U.S. Was Not So Far Away

The boat journey was long, and the United States was far away. But, to the Issei, the U.S. was not considered so distant as to be unreachable.

Japan is surrounded by the ocean which once served as a wall cutting off the country from the outside until the end of national isolation in 1853. Thereafter, the ocean became the route to expansion. The Japanese government made steady progress toward high industry and capital, thus strengthening the country. The victories of the Sino-Japanese War (1894-95) and the Russo-Japanese War (1904-05), the annexation of Korea to Japan (1910), and the Twenty-One Demands to China (1915) were the stepping stones of Japanese expansion to other areas in Asia until the end of the Second World War. The Japanese population was slowly moving out of the four main islands.

Of all the prefectures of Japan, Hiroshima, Yamaguchi, Wakayama, and Fukuoka Prefectures contributed the greatest number of emigrants (Yoshida, 1909). Historically, Northern Kyushu (including Fukuoka) and the Sanyō region (including Yamaguchi and Hiroshima Prefectures) have been open to foreign culture and were places of foreign commerce, as they are geographically close to Korea and China. Wakayama Prefecture is mostly composed of mountains of thick forests which isolate it from the cultural centers to the north, Kyoto and Osaka. For a long time, the ocean was the main way to leave that area. Expansion by ocean has been a tradition in these districts for centuries.

In Watsonville in 1922, out of 136 heads of families from 20 prefectures listed in the *Japanese Who's Who in America* (1922), Issei males from Yamaguchi Prefecture numbered 35; Wakayama, 23; Hiroshima, 16; Kumamoto, 15; and Fukuoka, 15. The Japanese population was listed as a little less than 600 in the *Who's Who*, far less than the actual population at that time.[1] Although not representative of the total Japanese in the area, it shows the interesting trend that 25 out of 35 men who came from Yamaguchi Prefecture were from the same county, Kuga-gun. Similarly, 14 out of 23 men from Wakayama Prefecture also came from the same county, Higashi Muro-gun; 8 out of 16 men from Hiroshima Prefecture came from Aki-gun; 5 out of 15 men from Fukuoka Prefecture came from Asakura-gun; and 8 out of 15 men from Kumamoto Prefecture came from Kamoto-gun. Before leaving Japan, many of these men apparently knew someone who had already come to the U.S.; so this country was not a totally unfamiliar, distant place.

They Were Not Asian Coolies

It is generally said that the Issei were poor farmers who came to the U.S. simply to make money. Yosaburo Yoshida stated, "A large proportion of the Japanese emigration comes from the peasant class in the districts of the south." He continued, "The inducements and attractions... are the result of the simple fact that labor earns more in America than in Japan" (op. cit., p. 164).

However, the Issei were not Asian coolies who merely worked like slaves for money. They had their own dreams and reasons for coming to the U.S. Yamato Ichihashi, in *Japanese in the United States* (1932, p. 82), noted the intelligence and ambition of the Issei. After comparing the Issei with other immigrants, he concluded:

> The Japanese immigrants exhibited a satisfactory average with respect to money in their possession, ability to read and write, and degree of intelligence and ambition.

The Issei are not as illiterate in English as is commonly thought.

Most, especially the men, can speak English well enough to communicate, at least on a rudimentary level. Living in the U.S. for many years, the men had to speak English to conduct business, and they gradually developed their language skills. The reports of the Immigration Commission (1911b, p. 443) describe the literacy of the Issei in the Pajaro Valley around 1910, though the number of Japanese from whom they received data was limited:

> Most of the Japanese men on the 20 farms from which data were obtained are literate. In fact, of 26 males, all but one could read and write his native language.... Of the 26 males, 10 could speak, read and write English. The remaining 16, with one exception, could speak English but could not read or write it.

According to the same report, the women, still very new to this country and having fewer chances related with people outside the Japanese community, showed very low literacy. The report states, "Of the 15 married women, on the other hand, 8 could neither read nor write [their native language]."

While there were economic reasons why the Issei chose to come to the U.S., interviews with the Watsonville Issei suggest that ambition, not mere economic pressures, lured them to this country.

Ambition

The United States was a rich country where the Issei hoped for a better life and economic status. Though life in this country was not easy, they welcomed the risk and adventure. Despite hardships, their hopes and dreams sustained their strength. The individual stories of the immigrants who contributed to the advancements in agriculture are full of their strenuous efforts under severe conditions. Knowing their background and how they came to the U.S. may give clues to understanding the backbone of their strength.

Toshi Murata is from a small farm village in Wakayama Prefecture. The village was poor, and before his father came to the U.S. in 1906, five or six men had already left the village for the faraway country across the Pacific. Even though his father left to earn money in the U.S., his mother never left home and took care of eight tans of farmland (one tan = 991.7 sq.m.). She was responsible for the family and received money from his father only occasionally. Two of his older sisters and two older brothers went to the U.S. In 1921, when Toshi, the youngest child, became 14, he was also called to the U.S. by his father and went to Castroville, approximately 11 miles south of Watsonville. He thought of his mother who had worked hard while his father was gone. "Beyond anything else, I thought of gaining bread."

He helped his father in the sugar beet fields until his father returned to Japan with some success in 1923. Toshi comments on the first time he saw the farm his father managed:

> It was beyond my imagination. He [Toshi's father] was using a huge tractor and ten horses. I think it was about 250 acres.

Frank Sakata, a Nisei, tells me about his father, Kyuzaburo Sakata, who was known for his talent and success in business. The village in Hidaka-gun in Wakayama Prefecture where his father was born was poor. "Since there was not enough to make a living by farming, they had to supplement it by fishing." His father finished six years of education in Japan and left Japan in 1900.

> There were already several men who had gone to Steveston, British Columbia, from my father's village before he left for Canada. . . . [He] was the oldest son among three brothers and two sisters. When he left Japan, he knew that the family responsibility as the oldest son went to his next younger brother. He had no intention of returning home to Japan, but wanted to make a living here in the United States. . . . He did not have much knowledge about the U.S. He kept asking his mother to let him go to Canada. His parents finally gave in when he was 15 years old, since his uncle was working in Steveston.

Unosuke Shikuma, one of the memorable names in Watsonville, was active in the Issei community before and after the war. He was born April 13, 1884, in Kuga-gun in Yamaguchi Prefecture and was the son of a country store owner. It was the only store in the village where every kind of goods was sold. His daughter-in-law, Hiroko Shikuma, explains:

> [He wanted to have] an additional drug store. When he came back to Japan, he thought of a department store, an American-style department store. It was his dream when he first came.

Unosuke Shikuma arrived in San Francisco in 1902 when he was 18 years old.

Some Issei who came to the U.S. were not small farmers. Yuki Torigoe's husband, Bunkichi, was skillful with his hands and managed a watch and bicycle store in the town of Watsonville for many years from 1909. He was born in 1884 and finished eight years of education in Okayama Prefecture. The family used to own a rather large patch of farmland and managed a wholesale store. Mrs. Torigoe remembers:

> Yes, he was the first son. His mother was always weak and spent their fortune away, which used to be quite a lot. When she became sick, she needed money for a doctor. Whenever she visited the doctor, she hired the jinrikisha [rickshaw] each time. She was raised as a daughter of a rich family. In case of need, she just rode on the jinrikisha to the doctor. One tan of field could be sold at 60 yen at that time, and she could spend 60 yen real soon. She had to pay the doctor as well as those people. They sold the land one parcel after another.... After all, he came to the United States because he was poor.

Shuro and Masa Kobayashi are among the few surviving Issei in Watsonville. Shuro's family owned a dry goods store in Japan, and her family were landowners in Ibaragi Prefecture. Both of them were well educated at that time, receiving certificates from high school. Shuro Kobayashi explains why he came to the U.S.:

> People whose parents were rich and who were very smart went to upper school. There were not many specialized schools at that time. There was Taisei Gakusha, a medical school in Hongo district. At this school, you could become a doctor if you took just two examinations. It was so quick to become a doctor that my brother persuaded me to go to this school. But I did not want to go to Taisei Gakusha at all.... I probably wanted to satisfy my vanity. At that time, there was a great encouragement to go abroad. I wanted to come to the United States.

Two of his older brothers also came to the U.S. His father was a businessman who made a connection with local tea farmers to bring tea to Yokohama and Tokyo for export at a time when there were not many steamships.

> My father was such a different person. When he was young, he managed the business like that, too. A relative who was born at my place became a policeman in Kōzushima in Kanagawa Prefecture. He brought mulberry trees back. He tried to bring them back by a small

> boat from Kōzushima, about 100 miles away. But when he came around some place in Izu – I do not remember the name – he faced a storm. His boat almost sank into the water, as it was such a small boat. He had no choice but to return to the watercourse [where he started]. He did not give up and tried again after waiting for the wind to stop. He again had to face the storm. He thought it must be the end of his life. He tied his body to the boat to wait for death. He was such a different person. [Laughter.] I took after his spirit and came to the United States.

Shuro Kobayashi came to the U.S. to study in 1907, receiving a student passport when he was 20 years old.

> I did not [have any savings with me]. I made a promise with a prefectural official that I would go to school and receive 100 dollars a month when I came to the United States. But that was only for the sake of formality. [Laughter.] I did not receive anything [from my father].

Later, he comments, "when I came to the United States, I felt like I was engaging in speculation."

Like most other Issei, he was unable to continue his education, though he tried and really wanted to go to school.

> I have never been a schoolboy, though I wanted to be. There was no opportunity for me. That was my hope and plan to go to school in the United States, but there was no opportunity for me.

Kumajiro Murakami, a pioneer Issei, was 97 years old when he was first interviewed in 1978. He recalls his memories of Japan:

> Young men were thinking about how to make money. Saga village faced four ri (one ri = 3.9 km) of the Bay. From above the hill, I once saw seven American combined fleets below our village. The Navy military school was four miles away.

Kamiseki in Saga, Kumage-gun, has been an important port for many years.

To know the world outside the village was the pride of young people and a way to gain recognition. Murakami describes his decision to go to Hawaii:

> There were no particular reasons.... There were many people who went to Hawaii from my village. Five or six men got together and

went to Hawaii. They were showing off their boots for the trip to Hawaii, though it was not necessary to wear them. It looked good to me, and I joined them.

In the country, school was far away, and children often had to walk over a mountain pass to get to school. Fuji Murakami, his wife, grew up in a nearby village and could not continue her education.

I went to school for a short time. But my school was such a long distance away. Four boys from Waki-mura stuck up their noses and taunted me whenever I was on the street going to school by myself. I did not like it, so I quit school early. . . . At that time, you did not need to go to school that long.

Kumajiro Murakami also stopped going to school when, "just at the time of my fourth year in school, my ear started to ache." He was 17 years old when he left Kumage-gun in Yamaguchi Prefecture in 1898. He recalls his journey on the boat:

Because of the bad water, I got terrible diarrhea. I had to go to the bathroom every hour. It is not clean talk, but I had to stay at the edge of the boat and do it. When we arrived in Hawaii, many Japanese celebrated our arrival with plenty of food. I could not eat at all. I was able to recover from the diarrhea because I drank the fresh blood of a chicken.

It was a surprising experience to go to Hawaii. . . . In Hawaii, there was a boss from Yamaguchi Prefecture who could speak English. The house was simple, five to ten men slept in the same house with just their blankets on the wooden floor. It was not cold at all in Hawaii.

I got a 106° fever in Hawaii. It was a miracle that I survived. There were about 60 people in the hospital, and about one or two people passed away daily due to the fever. There were too many people to take care of, but the cook, Mr. Shimizu from Hiroshima, gave me rice broth instead of just water. He cooked rice with plenty of water and cooled it off in the icebox. I learned about the essence of rice from that experience. Because of the rice broth, I was able to recover from the fever. Most of the patients were Japanese. I do not remember exactly when it was, but it was just before I moved to the mainland.

In Hawaii, "the more you harvested, the more you got paid." Unfortunately, it did not rain and there was no harvest. After working at sugar cane with little payment for many years in

Hawaii, Murakami decided to move to California, since he knew people who were working on strawberry farms in Watsonville. He was the first among his friends from the village to go to the mainland.

With a Blanket over My Shoulder (Migratory Labor)

The Valley Is Rich

> Standing on an elevation of the Santa Cruz Mountains and looking out toward where the horizon dips in the Pacific, one perceives the Pajaro Valley in all its beauty of form and dress. Near the foothills is a chain of beautiful lakes, and useful too in that they furnish water for irrigating berry fields and garden products requiring an excess of moisture. Beyond them one sees the varicolored fields of wheat, barley, corn, hops, beets, potatoes and beans, orchards and vineyards, presenting the appearance of a wonderful mosaic (Harrison, op. cit., p. 20).

Before the turn of the century, California was a bustling state still in the process of being settled. According to the 1880 census, Santa Cruz County had a U.S. native population of 9,655, only 5,792 of whom were born in the state of California. The foreign-born population totaled 3,147, about one-fourth of the total population. The first appearance of Japanese in Santa Cruz County was as early as 1886, according to Karl Yoneda (1971, p. 151):

> [T]wo Englishmen named Amore and Goldman brought 9 Japanese farmers to Soquel near Santa Cruz to plant tangerines. This failed. They then moved on to Woodland to operate a farm; however, the Englishmen abandoned the men because of financial difficulties.

Japanese Club early 1900s in Watsonville; courtesy of Shiryoshitsu/California

First Bank/S.F.; reproduced by Chris Huie

Sometime around 1892 the Japanese first moved into the Pajaro Valley. A contractor brought two dozen men as seasonal laborers, Surprisingly, Sakuzo Kimura, the first contractor, started his profession very smoothly. It helped that he once worked for the U.S. Navy and was fluent in English. He was around 40 years old, very active, and managed to make contracts, one after another, at a sawmill ranch in Aptos and at a 30-acre hops farm off East Lake Avenue in Watsonville (see Kato, 1961, p. 421).

Kimura moved into the Valley at the right time. Strawberry cultivation was already well developed, increasing in acreage to 552 in 1895 (*California Cultivator*, 1922) and supported by Chinese workers. Japanese participation as laborers in the strawberry fields started around this time. Until the turn of the century, most Japanese worked in the sugar beet fields. Claus Spreckels, who made a fortune out of sugar in Hawaii, moved to Pajaro Valley and built a sugar factory in 1889. According to the California State Board of Agriculture (1913, p. 840), many experiments in the sugar beet industry in the U.S. ended in failure. "With the building of a factory in Watsonville [in Monterey County] by Claus Spreckels, and others by the Oxnard Bros., the industry rapidly developed."

The same report indicated how much California sugar beet production increased: 5,170,350 pounds in 1889; 9,250,200 pounds in 1890; jumping to 21,801,330 pounds in 1893; and double that number the next year. There was a great demand for laborers, especially in the Valley where the Spreckels sugar factory was built.

At first, Spreckels bought sugar beets from independent local farmers. He later sought to acquire total control over the farmers' production: "[T]he sugar-beet factories at first encouraged small farmers to raise sugar beets under contract" which could lead to a "concentration and monopoly of business" (McWilliams, op. cit., p. 84). But this had yet to happen in the Valley.

Japanese Learn the 'Boss' System from Chinese

It is said that the Japanese learned their "boss" system from the Chinese. In California, the Chinese has been the labor force in the fields before the Japanese came.

[T]he Japanese . . . soon adapted themselves so as to fit into the Chinese system of labor and living conditions. Among other things, they adopted the Chinese "boss" system (U.S. Immigration Commission, 1911a, p. 62).

In the Pajaro Valley, "They [Chinese and Japanese] have worked under the same 'boss' system and have lived on the ranches under the same conditions" (U.S. Immigration Commission, 1911b, p. 442). They were of the same Asian race and were treated differently from the white workers.

[Chinese] were living in the valley, thinning, hoeing, and harvesting sugar beets, growing hops, and doing the greater part of the agricultural work not done by the owners of the land and their few regular white employees (ibid., p. 432).

The bosses did not simply supply labor, but they also had to understand the business of farming. The *Watsonville Pajaronian*, on July 26, 1894, commented on Chinese bosses:

The Chinese bosses are good judges of the coming beet crop, and they all say that the coming crop will be a mammoth, and that 20 tons to the acre will be frequently reported.

They were right. The sugar beet crop in 1894 was very good in the Valley.

The reported Chinese population in Santa Cruz County was 6 in 1860; 156 in 1870; 523 in 1880; 785 in 1890; and 641 in 1900. The Chinese had supported Valley agriculture for a long time. Among the various jobs the Chinese did, E. S. Harrison (op. cit., p. 30) mentioned strawberry cultivation at the Lake Farm around 1890:

The method of conducting this farm is to let it to Chinamen upon shares, the owner furnishing everything, the Chinamen performing the labor and receiving one-half. On Lake Farm during the busy season from three to four hundred Chinamen are employed, while in the winter about sixty are kept busy. These Chinamen represent a number of companies, and work under bosses. They live by themselves, mixing with the Caucasians only to the extent which necessity compels them in business relations.

Sakuzo Kimura established his own labor club in December 1893 (see Kashiwamura, 1911, p. 292). Though the clubhouse burned down in August 1897, he rebuilt the structure and named it the "Shinyu" (good friends) Club. Members would pay an annual fee, enabling them to receive mediation in the work, contracts, land leases, and other related business intercessions from the boss. Ichihashi (op. cit., p. 173) added that a Japanese man could pay annual dues of $3.00, "for which he was allowed to cook his meals or lodge at the club or both whenever he was out of work." By the time Kumajiro Murakami came to Watsonville around 1903, the club served meals, allowing the boss a small profit. His labor club fulfilled the needs of people from both sides – Japanese and local farmers – in the Watsonville area.

> In time this club became a general rendezvous for the Japanese in the district, and when employers needed extra hands they went to the club and secured the men they wanted (ibid.).

In general, the Japanese boss system seemed to function the same way as the Chinese boss system: as a mediator between laborers and employers, and as a center for their own people. It is worth commenting that Sakuzo Kimura, an important boss who was known to people in the Watsonville community, passed away on Sunday, May 13, 1900. The *Watsonville Pajaronian* (May 17, 1900) printed a small article concerning his death:

> S. Kimura, the pioneer Japanese contractor of this valley, died Sunday evening and was buried Tuesday. His death was due to consumption. Kimura was an intelligent Japanese, and had seen much of this world. He was in the U.S. Navy several years.[2]

Japanese Become Noticeable in the Strawberry Fields

In the Pajaro Valley in 1901, Spreckels could no longer enjoy the sugar monopoly it had held for so long, facing fierce competition from the American Sugar Refining Company (an Eastern trust controlled by Henry O. Havemyer).

> Spreckles [sic] refused to pay the local growers the initial price of four dollars a ton for beets. As a consequence, the growers ceased to grow beets around Watsonville (McWilliams, op. cit., p. 83).

At that time, several Japanese bosses had been contracting with

Spreckels in Monterey County. They were not outside this dispute and had close contact with Watsonville.

> About sixty Japanese laborers quit in a body on the Spreckels ranch at King City [in Salinas Valley] the other day. They had been acting in such a careless and independent manner that the foreman was repeatedly compelled to remonstrate with them, and at last was forced to lay off eight of the leaders in the disturbing element. Whereupon the whole gang walked off the field, remarking that Watsonville was more to their liking anyway, for the work was lighter, the sun wasn't so hot, and a Jap was looked up to more (*Watsonville Pajaronian*, June 13, 1901).

One Japanese publication[3] indicated that workers were paid from $1.00 to $1.50 per ton of sugar beets in 1901 and were told to lower that amount in 1902. Spreckels moved to Salinas in 1902 and started its sugar trust company again.

Spreckels did not stay long enough to gain control over the local farmers and make the Valley into one expanse of sugar beets. The Pajaro Valley generally remained in small holdings by individuals. Small-scale individual farms allowed laborers personal contracts with the employers in contrast to large-scale industrial farms, where they would only know their employers by their names on the contract.

The Spreckels' move was a significant event for many Issei in Pajaro Valley. Sugar beet was the most important crop in the Valley and provided the most jobs for Issei. It not only helped the development of labor clubs but also enabled a few Issei to save enough capital for future businesses. Above all, as Japanese participation in sugar beets in the Watsonville area declined, their participation in the berry fields increased.

It was rather easy to start the contract leases for strawberry farms since the owners provided everything and the Japanese provided only the labor. Work in the berry fields was much lighter than work in sugar beet fields. Yamato Ichihashi (op. cit., p. 179) lists the payment to Japanese who made a contract lease for the strawberry farms in Watsonville for 1908-09:

First year, setting plants, per acre	$20.00
Second year, watering, weeding, hoeing, picking, packing, and loading, per acre	$50.00

Third year, the same work, per chest	$ 1.50
Fourth and fifth years, per chest	$ 1.75

Those leases provided quite different circumstances to Japanese as the contract lasted for four to six years. Instead of picking and packing as short, temporary laborers, Japanese settled in one place for a longer period of time. Soon, Japanese became very noticeable in the strawberry fields. The acreage of the strawberry cultivation in the Valley was 42 acres in 1881, 268 in 1885 (Wilhelm and Sagen, op. cit., p. 176). Another source indicated, "The planting [of the strawberry] increased until in 1895 there were 522 acres, and it remained about the same until 1900. In 1901 there were 700 acres; 1902, 840; 1903, 800; 1904, 700; and 1905, 670" (*California Cultivator*, 1922).

On Their Own Two Feet

In 1900, the entire foreign-born population was 4,979 out of a total population of 21,512 in Santa Cruz County. Immigrants from many nations crowded themselves into the county, searching for jobs and settlement: 628 from Germany, 596 from China, 513 from Ireland, 460 from Italy, and many more from 30 other countries. Japanese were among those increasing the population: 19 in 1890 and 235 in 1900. Many of those immigrants, working up to 10 hours a day, supported the development of the Valley's agriculture.

Young bachelors hoped to earn money, marry, and settle down, leaving the ranks of the migratory labor force. They tried very hard to be economically and socially independent, creating "homes" of their own in this country.

Among these immigrants, the Dalmations found a way to climb up the economic ladder by specializing in apples. The Dalmations had a special ability to examine buds, so that they purchased the first crops on the branches at budding time after careful examination of each blossom. By 1910, "about one-third of the Dalmations with families own or lease farms" (U.S. Immigration Commission, 1911b, p. 433).

The Immigration Commission (ibid., p. 432) commented, "It must be said, too, that the influx of the Japanese has had something to do with the elimination of the Chinese, since members of that race competed with the Chinese for the hard work to be done." However, it was the Chinese Exclusion Act of 1882 that cut the flow of Chinese immigration to the U.S. Also, most Chinese retired from the fields as they got older to settle down to managing independent businesses. The same report says that Chinese "operate and 'man' two small establishments for evaporating apples and engage to a slight extent in small farming."

The Chinese retired from the fields sometime between 1900 and 1910, when Japanese showed great progress in their strawberry cultivation in the Valley. At this time, the Japanese population surpassed the Chinese population in Santa Cruz County. The U.S. census indicated that while the Chinese population in Santa Cruz County diminished (745 in 1890; 614 in 1900; and 194 in 1910), the Japanese increased (19 in 1890; 235 in 1900; to 689 in 1910). As early as June 1894, the *Watsonville Pajaronian* noted:

> The Japanese are becoming more and more plentiful in this valley, and in certain classes of work seem to be crowding out the Chinese.

Cutting Open Their Paths

It was a dream of the Issei to succeed in this country. They were young and resilient enough in spirit to overcome the difficulties they had to face in the beginning.

Shuro Kobayashi's older brother went to Vancouver as a tea trader with a merchant visa.

> My brother became a fisherman north of Vancouver. He brought fish to send to Japan. He was a merchant. I mean, he had the spirit of a merchant. Beyond his expectation, it was not a profitable business. He quit as a fisherman and came down to the United States.

Shuro Kobayashi arrived in Victoria, B.C., in 1907 and traveled to many areas along the Pacific coast for work. After one year of traveling and being unable to become a student, he settled in Imperial County, California, as his brother had moved to Brawley to farm.

In Steveston, Kyuzaburo Sakata went to a missionary school to study English, so he could speak just enough to communicate with people. After staying with his uncle for two years, he decided to move to Lompoc in Santa Barbara County. Frank Sakata, his son, explains:

His uncle was always broke since he was constantly drunk, loved to gamble, and visited girls. He had to advance money to his uncle many times. He knew he would not be able to save much money. He wrote a letter to his fellow Kenjinkai [a group formed by people from the same prefecture] in Lompoc since he knew three of them from the same village. They encouraged him to come down. He came to Canada around 1900 and came down to the United States in 1902. In Lompoc, he was a common farm laborer at first. Later, he became a partner with two of his cousins, his only two relatives in the United States, and farmed with them on a share arrangement.

After staying in San Francisco briefly, Unosuke Shikuma moved to Watsonville, where he attended a mission school to learn English. In 1907, his son, Mack Shikuma, was born in Pajaro in Monterey County, where his father was working as a seasonal laborer. Mack's wife remembers:

> I heard that your father went to pick plums in San Joaquin Valley, and [when he came back] he was surprised to see how much you grew.

Mack also remembers that his father traveled around for jobs. But the next year, in 1908, Unosuke Shikuma was a member of J. S. Kōsansha, a company which grew strawberries and other crops on large acres of leased land.

When Bunkichi Torigoe arrived in Vancouver, he was only 15 years old. He soon moved to Seattle and moved again to North Dakota to work for a railroad. His wife explains:

> There were many Japanese working over there. He was only 17 or 18 years old at that time. Since he could speak English, he became a translator for everybody. He could not do any hard work at the railroad. But he was such a bright man that once taught he could direct the connecting of tracks without any difficulty at all. He was doing such an easy job.

She must have heard the same story from him many times. She speaks as if she experienced it herself:

> At that time, it was just like a movie. There were gambling houses and gangs. Everybody carried pistols on their waists. . . . Cowboys came to shoot Japanese in the camp. Papa was just so surprised to see this, his eyes popped open. [Laughter.] They did such crazy things.

After staying in North Dakota a little over one year, he moved to San Francisco, where he worked for a white family as a cook's helper. Then he moved to the Watsonville area.

Kumajiro Murakami decided to go to the mainland from Hawaii since he knew some farmers in Watsonville.

> I wanted to go to the U.S. I wrote a letter to the Japanese inn to see whether I could go to the mainland or not. They responded that it was quite possible, and to come down as soon as possible. I had a farewell party, drinking one cask of sake. I was on the small boat to Honolulu. I could see people waving their arms and clothes at me on the hill.

He arrived in San Francisco around 1903.

> There were many people who went to the railroad in the east after arriving in San Francisco. I went to Watsonville all by myself. I could speak English well enough to ask the directions. In Hawaii, I went to visit the Englishman's wife to learn English a few times a week. The school was two and a half miles away from the camphouse. I was the only one who went there. That was why I was better at English compared with others. I could come down to Watsonville all by myself though no one was waiting for me.

Later, he also became a member of Y. Kōsansha and grew strawberries on large acres of leased land.

Labor Clubs

The Issei showed great desire to be independent farmers, though most lacked the capital. Therefore, they looked for ways to achieve independence. The boss system was one way for Japanese to progress, and they developed it to suit their particular needs.

> Whereas the Chinese gangs were simply a form of organization of cheap labor convenient and profitable for the employer and the Chinese contractor, for the Japanese the contract system became the central *instrument* of a rise from coolie status through sharecropper, renter, manager, owner, and finally to a position of virtual monopoly in certain specialized vegetable and berry crops (Fisher, op. cit., p. 25; emphasis mine).

The Issei were not bound to the labor club. They only stayed in a club as long as they wanted to accept the club's services.

By 1910, the labor clubs were well established; membership was open to everybody who paid an annual fee. Assistance in obtaining a single contract was also provided for a commission of 5 percent of the daily wage. The clubs controlled the members by negotiating contracts with employers. The clubs could be powerful; as the Immigration Commission commented (1911b, p. 435), determination of wages was "absolutely controlled by the members of these clubs."

The clubs did not simply help the Japanese find jobs but managed with great efficiency to provide news about jobs for migratory laborers. They exchanged such news with bosses in areas beyond the Valley.

> When the demand for men was more than the members could supply, he sought outsiders.... When the season of the district began to slacken, the "outsiders" first withdrew, and some of its members also migrated whenever it was found advantageous to work elsewhere. To assist these migratory members, the secretary studied the situations in the neighboring districts if he did not know them already (Ichihashi, op. cit., p. 173).

They provided needed hands to every crop in the Valley and expanded their services to neighboring counties.

Importantly, their management was not limited to contract labor. They handled the process of leasing land as well as houses. Members paid 2 percent of harvested crops with the membership fee if they were the tenants of the labor club with a lease (Kashiwamura, op. cit., p. 291). The Shinyu Club even handled the processing of paper for civil and criminal lawsuits, advertising that service in the *Japanese American Yearbook* in 1910.

Rikimatsu Tao, from Aki-gun in Hiroshima Prefecture, ran the Shinyu Club. This became the most popular club with about 200 members around 1910. Kumajiro Murakami remembers: "Mr. Tao was a feisty man. He would answer back, 'I will give you my fist,' whenever somebody did not listen to him."

Tetsutaro Higashi, from Kuga-gun in Yamaguchi Prefecture, managed the Nichibei (Japan/America) Club which split from the Shinyu Club once he became the boss. His English was good, according to Kumajiro Murakami, and his club's advertisement in the *Japanese American Yearbook* was in English:

The Japanese American Employment Club
all kinds of labor furnished
all members of our club are expert workers
T. Higashi
[manager]

Both Tao and Higashi came to the U.S. very early, Tao in 1893 at the age of 24 or 25, and Higashi in 1892 at the age of 33 or 34. They were older than the other Japanese around, becoming long-term residents in the Watsonville area. They learned leadership by assisting Sakuzo Kimura, an original boss of the Shinyu Club until his death in 1900. Many people from Hiroshima gathered around the Shinyu Club, and people from Yamaguchi Prefecture gathered at the Nichibei Club.

There were more clubs established in the Watsonville area around 1910 (ibid., pp. 292–293). The Kyoeki (common benefit) Club was managed by Risaku Hirabayashi from 1904. In the same year the Nihon Club was established by Kōuemon Tanaka, but Itsuto Matsuoka took over the leadership in 1907. All the clubs had more than 100 members in their prime.

Japan Town

A Japanese town began to appear around 1905. At least five stores were established by 1906, as listed in the *Japanese Who's Who in America* (1922). Yasutaro Iwami from Kuga-gun in Yamaguchi Prefecture settled in Watsonville soon after coming to the U.S. in 1900 and opened a barber and billiard store. Keizo Atsumi from Toyohashi City settled down to open a tailor shop after coming to the U.S. in 1901. Nisaburo Madokoro from Higashi Muro-gun in Wakayama Prefecture opened a *manju* (Japanese sweet cake) store sometime after 1900. Tokuzo Oda from Hiroshima Prefecture opened a barber shop and public bath around 1905.

The names of more stores appeared in the local column of the *Shinsekai* newspaper published in San Francisco. In 1907, many other establishments were mentioned in this newspaper: Moriyasu grocery, Asaga shoe store, Murakami tofu factory, Fujii tailor, and "Kagetsu-do," a Western sweets store.

To serve the needs of bachelors, traditional Japanese restaurants offered Japanese food. The *Japanese American Yearbook* in 1910 listed "Yachiyo-tei," "Asahi-tei" and "Ichinofuji," all of which were in Chinatown across the Pajaro River. Yuki Torigoe remembers the one in the Japanese residential area:

> By the time I came [in 1914], there were Japanese restaurants. Maruman restaurant. There were waitresses; a few women always worked there. They [men] used to go there at night.

Many of them became longtime establishments of Watsonville, but others disappeared. One young man who managed "Kagetsudo" faced bankruptcy due to his excessive drinking and extravagant expenditures (*Shinsekai*, Aug. 19, 1907). Some might have returned to Japan.

A majority of the Issei were still transient contract laborers, and the labor clubs in Watsonville continued to assist them. The clubhouses were not only agents that looked for jobs but also places to which men could return whenever their contracts were over.

When they were not working, some of the Issei stayed at a boardinghouse in town. Japanese boardinghouses may have appeared sometime after 1900. Senkichi Gyōtoku from Fukuoka Prefecture opened up the boardinghouse "Daikokuya" during the Russo-Japanese War (1904-05) when more Japanese moved to Watsonville. Yonosuke Fukumoto, from Kuga-gun in Yamaguchi Prefecture, who had worked for Sakuzo Kimura since coming to the U.S. in 1892, opened Fukumoto Inn in 1906. By 1910, there were 10 boardinghouses. Like a clubhouse, the boardinghouse was not only a place to stay overnight but also functioned as an employment service. Kataoka boardinghouse advertised that it was recruiting railroad workers.

In order to sell goods to the Issei farmers scattered all over the countryside, peddlers traveled from labor camp to labor camp. Yamamoto Pharmacy peddled by walking. Yamamoto thought a bicycle was too much of a luxury item (*Shinsekai*, June 21, 1907). Gorokichi Takasugi borrowed a buggy daily to drive around the countryside selling fish (*Shinsekai*, Aug. 12, 1907).

In 1907, *Shinsekai* reported that two grocery companies had been established: Seno, Itō, Nishi, Hanamoto and Endō organized one in Castroville; and Hirai, Torigoe, Nakanishi and Sakamoto formed one in Salinas. The Salinas company was named "Hōkoku-shōkai" which began with $30,000 in capital and six hundred shares of stock worth $5.00 each (*Shinsekai*, Feb. 16, 1907). Two hundred shares of stock were owned by the members, but the rest were sold to the public. Moriyasu, Higashi, and Miyazaki in Watsonville became their stockholders.

Yuki Torigoe explains how Hōkoku-shōkai was managed:

> When he [her husband, Bunkichi Torigoe] was 21 years old, he participated in the grocery business, Hōkoku-shōkai in Salinas. He used to sell rice, soy sauce and other items by taking orders. He rode horses to deliver them to the country. He could not get paid for those because people in the country were not doing good.

By 1910, the number of businesses in town had boomed. The Immigration Commission reported 37 in 1909 excluding two doctors, and there were 50 listed in the *Japanese American Yearbook* and 51 in *Hokubei Tōsa Taikan* in 1910. The differences in number seemed to depend on how the businesses were counted. Several businesses were located at the same place under a single management. Businesses included groceries, boardinghouses, Japanese and Western restaurants, barber shops, billiard parlors, Japanese and Western bathhouses, watchmakers, photographers, a stagecoach company, tailors, a laundry, a shoemaker, a tofu factory, a bicycle shop, a sweet shop, and doctors.

Most stores were managed by a single merchant with a little help from other family members. Nippon-shōkai, one of the variety stores, was a company led by Kameichiro Inoue. Higo Inn on San Juan Road operated on a larger scale, combining an inn, a restaurant, and a public bath. Yokohama Laundry had such good business it hired three men, and altogether, five people worked to meet the demand (*Shinsekai*, Jan. 21, 1907). They still washed clothes by hand, limiting the number of clothes they could handle.

Atsumi Tailors used a Tanomoshiko [money pooling system] to sell clothes.

> A monthly Tanomoshiko will be set up for 12 people to participate in a group purchase. Each person must pay a $2.15 installment every month, and each person will take turns to buy more than $25 worth of tailored clothes (*Shinsekai*, July 3, 1907).

Yuki Torigoe explains how her husband started his watch and bicycle shop in Watsonville in 1909:

While he was managing the business in Salinas [Hōkoku-shōkai], he learned how to fix watches by himself. When he went to the town, he visited the watchmaker to ask how to repair them. He learned all about watchmaking by asking. He bought a bicycle to ride on and taught himself about bicycles.

His store was managed by his family without any other employees except his younger brother, who helped once in a while when he was not working. Yuki Torigoe continues: "Even the showcase he bought piece by piece and assembled himself, by looking at the order book."

With a Blanket over My Shoulder

Most people moved around, even to distant places wherever a job was available. Masao Kimoto, a Yobiyose Issei who came to the

Torigoe store from Torigoe collection, 1913
reproduced by Media Services/U.C.S.C.

U.S. in 1914, described the life of bachelor men. Many men traveled carrying the "buranketto" (blanket) over the shoulder.

> If one did not like something, one could just leave within minutes, rolling up one's bed in a blanket. The employer provided a bed, made out of wooden frames in which hay was laid. You could make your bed immediately by rolling your blanket out on the top of that bed.

The houses provided by the owner of the ranch were "bunk-houses of the cheapest sort" (Ichihashi, op. cit., p. 174). The life of the bachelors was easy-going and without responsibilities, but it was rough and crude. The Watsonville local column of *Shinsekai*, the Japanese newspaper published in San Francisco, listed several fights between Issei in 1907:

> Mori and Yamamoto fought each other with a knife and saw at the Sasamoto prostitution house. Yamamoto got a big cut on his neck, and Mori's face was slashed by the saw and he suffered internal bleeding. They settled their quarrel through a mediator, but the cause of their fight was the accusation of Yamamoto about Mori's abusive language over excessive drinking (*Shinsekai*, Jan. 23, 1907).

It was not clear exactly where this Sasamoto prostitution house was located. A number of restaurants, gambling, and prostitution houses were lined up on both sides of the street across the bridge in Monterey County. The street was not paved; whenever it rained, only boards laid down for passers-by protected boots from sinking into the mud.

"Across the bridge" meant Chinatown, where many Issei spent their meager savings for entertainment, especially gambling. "Ba-kappei" was one of the gambling games, similar to bingo. In this game Chinese characters were used instead of numbers. A Chinese American who grew up in Watsonville explained, "You just chose eight characters out of these 80 characters on the sheet, and if your choice of characters matched the one they chose, you won."

Gambling was popular among Japanese for many years. Not only single men, but married men, even those who had children, went to Chinatown to gamble. *Shinsekai* commented on the popularity of gambling and also mentioned women gamblers (Nov. 13, 1913). Instead of saving their money, they "gave money away to Chinese." A Chinese gambling house was thus called a "Shanghai Bank."

Even though the numbers of brides increased, many single men sought the company of the opposite sex in prostitution houses. Yuki Torigoe used to sit in her husband's watch and bicycle store on Main Street.

> I used to see Japanese women in high tone dress walk by our store. They were nice-looking young women. When young men called them, they would turn back and smile.... They dressed up in Western clothes, just like white ladies. They got a lot of money.

Shinsekai published several cases of "improper" women throughout the year of 1907:

> A wife of Inoue who works at Frank Eaton's farm in Pajaro had an affair with another man, though she had a seven-year-old boy. He got rid of the evil [wife] but did not know what to do with his child. It is decided that the boy will be taken care of by the Buddhist church (July 28, 1907).

Most Issei were single young men with high spirits. Shuro Kobayashi, who lived in Brawley at that time, says:

> People were full of pioneer spirit, different from people these days. They did not drive themselves hard. If they had spare time, they would go to Los Angeles to have fun.

Succeeding in This Country

Nevertheless, many Issei eventually put their hopes into succeeding in this country, pouring their energy into business. They worked hard all day, hoping to improve their status.

Many Issei became half-share strawberry farmers. In this arrangement, landowners and farmers shared the profits equally with the risk of the fluctuating market price, doing exactly the same work as the contract farmers. Japanese farmers could take risks which could increase their income, unlike those working under a contract for a fixed wage based on the lowest market price (Ichihashi, op. cit., p. 180).

The next step leading to independence and progress was to lease land as a cash tenant. Though it required a large sum of capital with which to start, leasing land offered greater benefits and freedom than the half-share system. According to the *Hokubei Tōsataikan*

(1911[4]), Ueda, a brother of Rikimatsu Tao, was the first to lease a strawberry farm in 1900. The next year, Nishimura and Higashi leased 30 acres of strawberry farms. They were among the few Issei who were able to lease land so early.

Around the time of the Russo-Japanese War, many other Issei pooled their money and were able to lease land. A group of men started to lease a large acreage of land for strawberries, a few years after Kumajiro Murakami came to Watsonville. The group was called Y. Kōsansha Company. He explained, "I was a replacement for my brother who was called back home to Japan due to our father's death. That was why I joined Kōsansha."

Y. Kōsansha was one of the Japanese farmers' companies and was formed with people from Yamaguchi Prefecture. Three Yamamoto brothers, Kumajiro Murakami, and Taroichi Tomita operated 50 acres of strawberry farmland before 1910. They had $12,000 in capital and paid $2,400 for leasing the land. They purchased a pumping machine with $650, dug a well with $300, and constructed elevated flumes to transport water for irrigation with $1,500. They constructed three houses and used four horses (Kashiwamura, op. cit., p. 294).

The same source indicated that J. S. Kōsansha was formed later in 1908 with people from Fukui Prefecture, except for Kōtaro Shikuma and his nephew Unosuke Shikuma, who were both from Yamaguchi Prefecture. Members included three men from the Kajioka family and Otokichi Yamaguchi. They leased 50 acres of land in 1908, later expanding the acreage to include crops besides strawberries. This Kōsansha fell apart early and was not memorable to many of the Issei with whom I talked.

Several Issei took the initiative to challenge themselves in this country. They were ambitious and willing to learn from one another. Otokichi Kajioka took a tour in Southern California to learn the state of farming there (*Shinsekai*, July 17, 1907).

With the establishment of farm companies, many men in Kōsansha were ready to get married. Otokichi Yamaguchi married the niece of Otokichi Kajioka in 1907. Later, Otokichi Yamaguchi changed his name from Otokichi to Masaki. Young Unosuke Shikuma married his cousin from Japan in 1906. By 1909, when Fuji Murakami came to their camp in Pajaro town, a number of women were already there. In 1910, there were 168 Japanese women in the

Pajaro Valley, 144 in rural areas and 24 in town. Children numbered 134 (U.S. Immigration Commission, 1911b, p. 436). Fuji Murakami remembers:

> I was brought to the Japanese camp soon after my arrival. I did not associate with any white women in the United States. It [the language] was not at all a problem. We were doing business in Pajaro. People from Yamaguchi Prefecture were already there.

The Murakami family grew strawberries until the Second World War broke out. Fuji Murakami, born in 1888, now walks with a stoop, commanding herself with "ohho ho, ohho ho" at each step. She says, "I used to pick strawberries and weeds." Weeding was needed constantly for the strawberries. Mrs. Murakami worked in the field for many years. "Your stomach always has to be filled," she says.

The harvest was hard work for everybody. A wooden chest (42″ wide, 17″ high and 17″ deep) with twenty drawers was used to pack strawberries. The chest was heavy and had to be returned to the growers. Kumajiro Murakami worked transporting strawberries. "It was really hard to carry them, since one chest weighed 175 kin

A chest at Shikuma residence
photo by Mona Nagai, 1983

(1 kin = 600 g). Three tons of load were carried by the wagon, pulled by horses." The weight of the chest varied, "A chest of berries is supposed to contain 100 pounds of fruit" (*Pacific Rural Press*, June 17, 1905, p. 370). The chest itself weighed about 50 pounds. It required heavy physical labor to carry a number of chests full of berries up to the wagons.

Every day was filled with hard work, and life in the country was basically one of survival. Everything was constructed by hand. In this environment, somehow the Issei managed to make homes of their own.

Ichiro Yamaguchi, the oldest son of Masaki Yamaguchi, was born in 1908 in Pajaro. "I was brought up in the strawberry patch." He remembers one incident from his childhood:

> They had this water running in flumes. Sometimes the water doesn't run anymore, so they go to see [check], and see them [these kids] sitting in the middle of it, stopping the water. [Laughter.]

These flumes were different from the elevated flumes used to transport water to the fields. This flume was shaped like a long wooden box, buried in the ground. From holes on both sides, water flowed to the field. Water could be made to flow quickly by placing a board on the surface and pressing down against the water.

In the camp, everybody spoke Japanese, and life seemed to be an extension of the Japanese countryside. They had a *kudo* for cooking, an *ofuro* for bath, and an outhouse for a toilet. An outhouse was made simply, and the toilet facility was even more simple. "They just cut a hole in the ground, that's all," says Yamaguchi.

A *kudo* was very simply made with an open mouth on the top and front, usually made out of concrete or earth, situated inside of the house. Women cooked meals on the *kudo* using twigs pruned from the orchards. Even though the floor around the *kudo* was earth, these simple cooking facilities often caused fires. Yamaguchi continues:

> [We] usually [had] two *kudos*, one for cooking rice and one for hot water. Most of them used a kerosene stove for cooking *okazu* [side dish] and some other types of cooking.

"Oh, *ofuro* is the best thing. I still miss it," two Issei commented during a casual conversation. The square wooden bathtub was separated by a tin board from the fire. One Issei told me that they used seaweed to patch up leaks in the wooden bathtub. They washed their bodies outside the tub and soaked themselves in hot water to relax and get warm. Ichiro Yamaguchi explains how it was used:

> Usually, it was a community *furo* [since] there were four camps there. They put the *furo* right in the middle of them. Each took turns. They had a bell there. When they got through, they rang the bell, so that they knew when somebody was through, you see. The next one who came first got a chance to get in.

Japanese Tenant Farmers

The Issei made great strides. According to the Immigration Commission (1911b, p. 437), in 1910, there were about 32 cash leases, 61 share leases, and 60 contract leases.

The *Japanese American Yearbook* has more detailed information. Around 1910, Kōtaro Miyazaki owned 16 acres of apple orchards. Fifty-three individuals and six pairs of partners worked as contract laborers on strawberry farms. Only Itsuto Matsuoka was listed as contracting 35 acres of sugar beet farm. Four single men and one two-man partnership worked for apples, vegetables, and other crops, but the rest (48 individuals, eight two-man partnerships, and one three-man partnership) were share tenants exclusively on strawberry farms.

Around 1910, the Issei began to settle down in the Valley, and strawberries had a great influence on this. "Most of the tenants have been married men who desired to have a settled residence for their families" (ibid., p. 437). Contract and share leases were easy ways to begin farming and did not require a large amount of capital, unlike cash tenancy. Besides, the land needed to raise strawberries was very small, usually less than 10 acres. The acreage of Japanese strawberry farms under contract and share could be as small as two acres, but most managed five to six acres.

Strawberry cultivation is such an intensive form of agriculture that one individual and a few helpers could survive from a farm as small as two acres. Small farms were easy to find, and Japanese farmers were welcomed by the local farmers, for not only were the Issei considered good workers, but also their hard work increased the value of the land.

According to the same *Yearbook* (1910), many Issei brought themselves up to the level of cash tenants. Of cash tenants, there were 19 individuals, seven two-man partners, three three-man partners, three or four large companies, including two Kōsansha, and one farm cooperative association. About half of these people grew strawberries, and the others tried various crops such as apples, onions, beans, vegetables, hay, and sugar beets.

It is hard to determine the exact population of the Issei in the Pajaro Valley. The *Japanese American Yearbook* lists about 200 Issei in 1910, a far smaller number than the actual population in Pajaro Valley. The adult male population at that time was about 500 in the rural area alone. Possibly there were more contract laborers and share and cash tenants, but many of the Issei were still transient seasonal laborers. Thus, some of the Issei might not have lived in the area continuously. Even the few Japanese who owned land did not always stay in one place. For example, in 1907, Chūtaro Terao sold six acres of land to Mr. Hiramura for $240 (*Shinsekai*, Aug. 22, 1907). After a certain amount of success, Terao might have returned to Japan.

Through the early part of the 1900s, the Issei in the Pajaro Valley were able to achieve farm tenancy. The Issei who cultivated strawberries usually remained on one farm for four to six years until the land was exhausted. Since it was easy to find new land within a close distance, the Issei could remain in a certain area for a long time. A community feeling gradually developed among them.

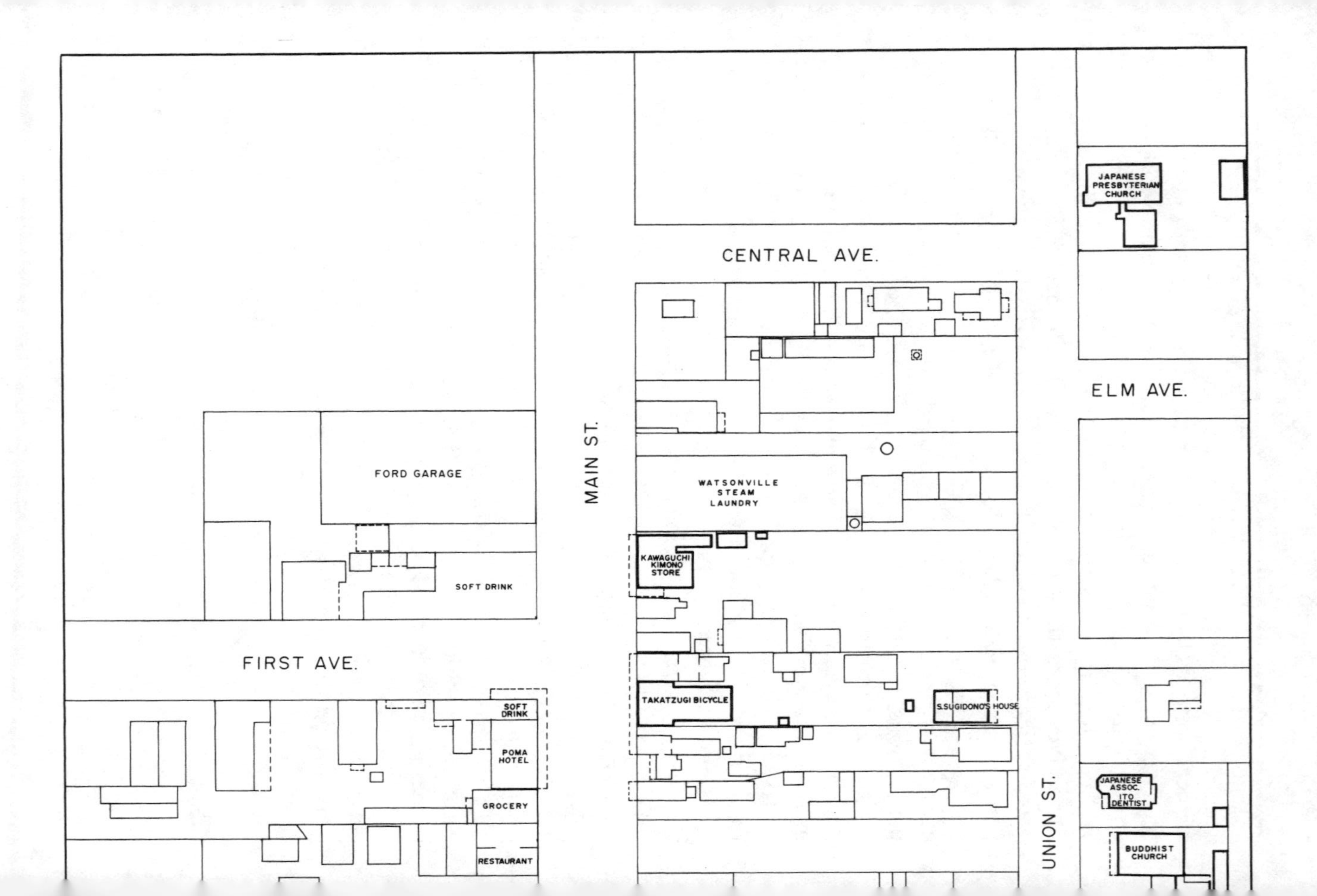
JAPANESE PRESBYTERIAN CHURCH
CENTRAL AVE.
ELM AVE.
MAIN ST.
FORD GARAGE
WATSONVILLE STEAM LAUNDRY
KAWAGUCHI KIMONO STORE
SOFT DRINK
FIRST AVE.
TAKATZUGI BICYCLE
S.SUGIDONO'S HOUSE
SOFT DRINK
POMA HOTEL
GROCERY
RESTAURANT
UNION ST.
JAPANESE ASSOC.
ITO DENTIST
BUDDHIST CHURCH

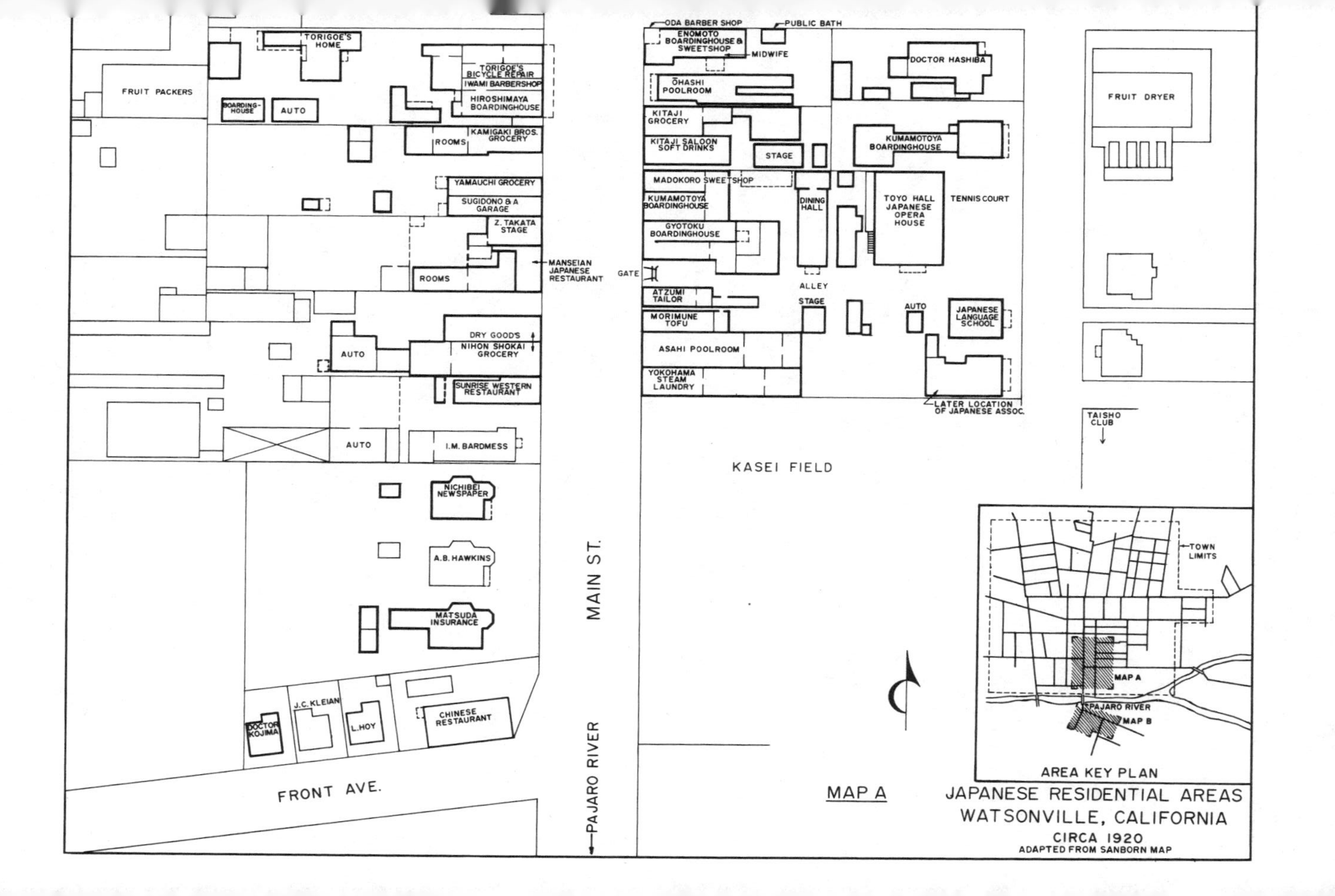

FRUIT PACKERS
TORIGOE'S HOME
BOARDING-HOUSE
AUTO
TORIGOE'S BICYCLE REPAIR
IWAMI BARBERSHOP
HIROSHIMAYA BOARDINGHOUSE
ROOMS
KAMIGAKI BROS. GROCERY
YAMAUCHI GROCERY
SUGIDONO & A GARAGE
Z. TAKATA STAGE
MANSEIAN JAPANESE RESTAURANT
ROOMS
DRY GOOD'S
NIHON SHOKAI GROCERY
AUTO
SUNRISE WESTERN RESTAURANT
AUTO
I.M. BARDMESS
NICHIBEI NEWSPAPER
A.B. HAWKINS
MATSUDA INSURANCE
DOCTOR KOJIMA
J.C. KLEIAN
L. HOY
CHINESE RESTAURANT
FRONT AVE.
MAIN ST.
PAJARO RIVER
ODA BARBER SHOP
PUBLIC BATH
ENOMOTO BOARDINGHOUSE & SWEETSHOP
MIDWIFE
OHASHI POOLROOM
KITAJI GROCERY
KITAJI SALOON SOFT DRINKS
STAGE
MADOKORO SWEETSHOP
KUMAMOTOYA BOARDINGHOUSE
GYOTOKU BOARDINGHOUSE
GATE
DINING HALL
ALLEY
STAGE
ATZUMI TAILOR
MORIMUNE TOFU
ASAHI POOLROOM
YOKOHAMA STEAM LAUNDRY
DOCTOR HASHIBA
KUMAMOTOYA BOARDINGHOUSE
TOYO HALL JAPANESE OPERA HOUSE
TENNIS COURT
AUTO
JAPANESE LANGUAGE SCHOOL
LATER LOCATION OF JAPANESE ASSOC.
FRUIT DRYER
TAISHO CLUB
KASEI FIELD
TOWN LIMITS
MAP A
PAJARO RIVER
MAP B
AREA KEY PLAN
MAP A
JAPANESE RESIDENTIAL AREAS
WATSONVILLE, CALIFORNIA
CIRCA 1920
ADAPTED FROM SANBORN MAP

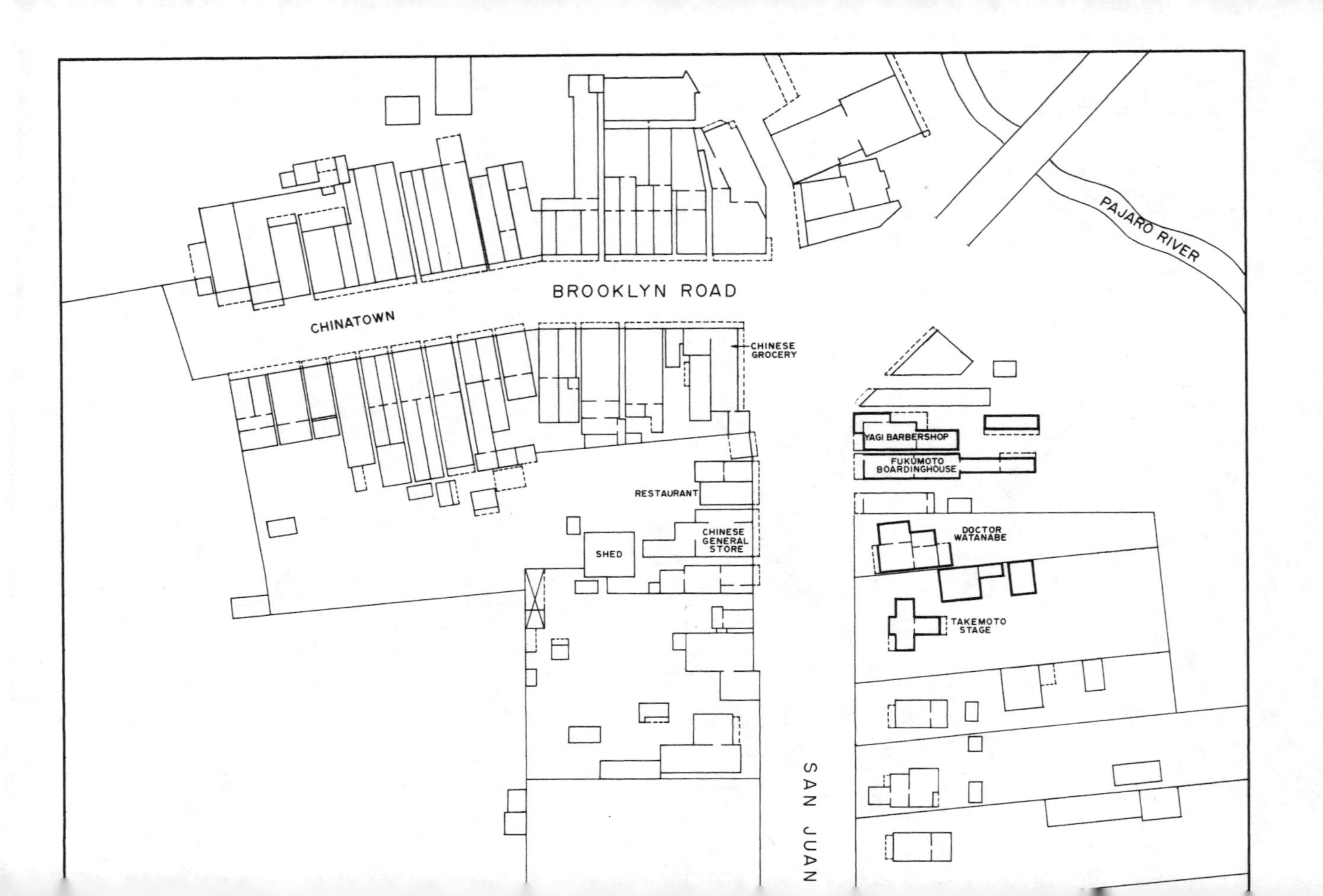
BROOKLYN ROAD
CHINATOWN
PAJARO RIVER
CHINESE GROCERY
YAGI BARBERSHOP
FUKUMOTO BOARDINGHOUSE
RESTAURANT
CHINESE GENERAL STORE
SHED
DOCTOR WATANABE
TAKEMOTO STAGE
SAN JUAN

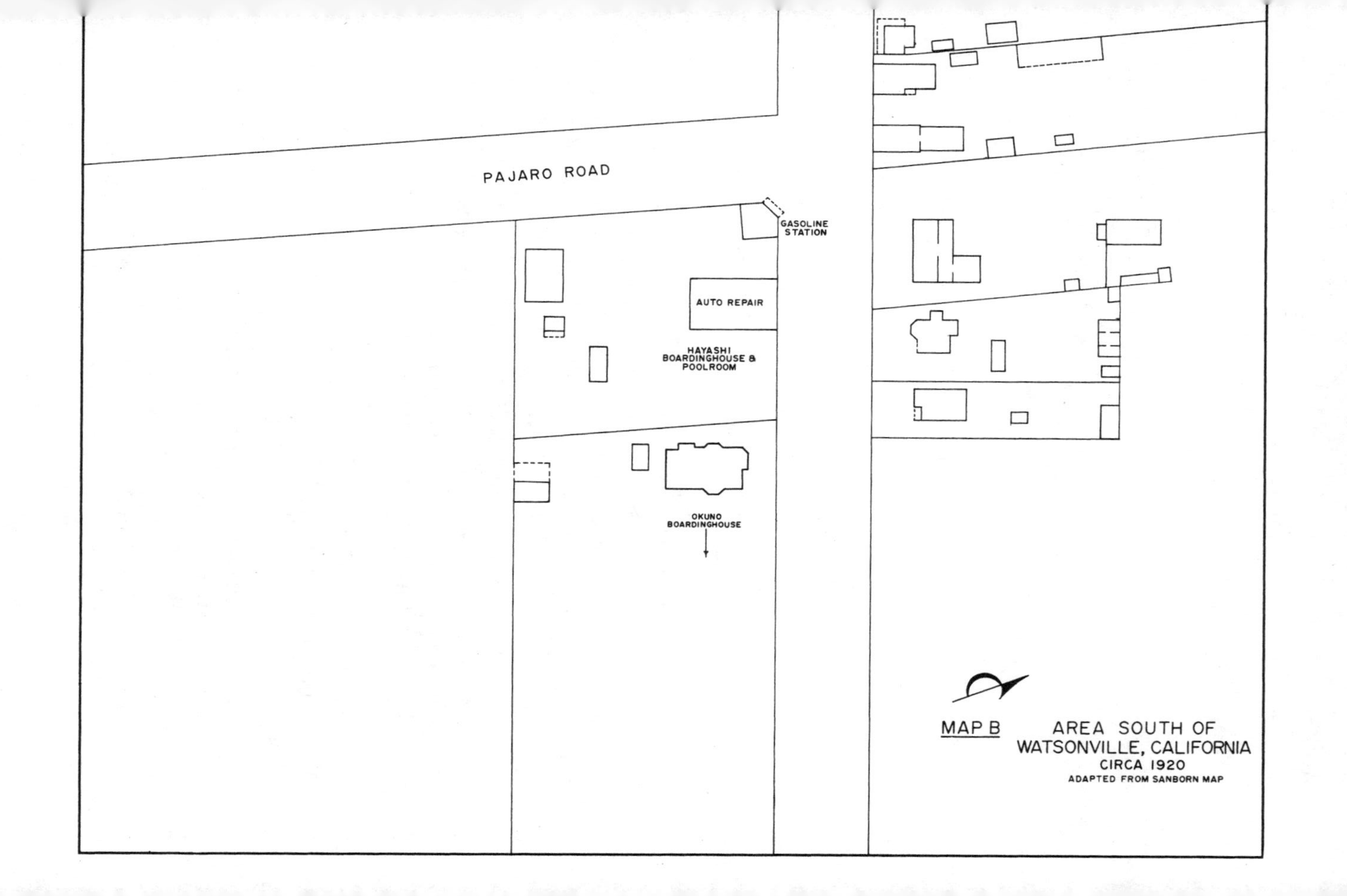

MAP B AREA SOUTH OF WATSONVILLE, CALIFORNIA CIRCA 1920
ADAPTED FROM SANBORN MAP

Families in a New Country

Starting a New Life

Watsonville is an old town. Chinatown was well developed before the Issei moved into town. The Japanese residential area was called "Nihonjin machi" (Japanese town) or simply "town" by the Issei. The Japanese area was located at the south end of Main Street, running through the middle of Watsonville. The area faced the Pajaro River and regularly flooded until a riverbank was reconstructed after World War II. The value of the land was very low. This area, degradingly called "lower Main," was right across from Chinatown, separated by the Pajaro River.

People who were once transient workers remember passing the area. An Issei who revisited Watsonville in February 1913 gave his impression of the town in the newspaper *Shinsekai*:

> When I look over Chinatown in the distance from that shaky wagon, I feel as if it were a desolate post town viewed from a train leaving the outskirts of Tokyo city.

But once he stepped into the middle of the town, he found the familiarity of the past as well as changes brought about by recent developments.

> There were many things that have changed, more accurately, things have developed. Mr. Shimauchi is very proud of his newborn Ameri-

can citizen. Mr. Akagi, a news reporter for the *Shinsekai* newspaper, sports a German-Yankee beard and looks splendid. There were innumerable things to comment on.... The street in the town used to look somehow dim and desolate, but it became bright under city lights. After all the bright city light is the life of night dreams and the culture of downtown.

Around 1905, as more Japanese migrated to the U.S. after the Russo-Japanese War, anti-Japanese sentiment became stronger and an anti-Japanese law was discussed in the California State Congress. It was also at this time that the Japanese in Watsonville began to establish their own organizations with the formation of churches and the Japanese Association.

Zaibei Nihonjin Kyōgikai (joint management of the community), who changed its name to the Japanese Association in 1909 (Kato, op cit., p. 422), was established to fight anti-Japanese laws. No formal Kenjinkai (a group formed by people from the same prefecture) was ever established, but religious organizations were important.

In 1898, a Japanese Christian mission was started by T. Terajima at his home on Main Street. After Terajima's death in 1899, the Rev. Ken'ichi Inazawa came from Salinas to take Mr. Terajima's place. Membership grew, and with the help of Dr. Ernest M. Sturge, a building site was purchased for a Japanese Presbyterian church at 214 Union around 1903 (Koga, 1977, p. 184).

The Japanese Presbyterian church provided an opportunity for Japanese to associate with local white people other than on a business level. A white woman provided an English class to young Japanese men. Unosuke Shikuma, who later became a devoted Christian in the community, took some of those classes. *Shinsekai* (May 2, 1907) describes one farewell party for two Japanese men, which Dr. Sturge and many other white people attended.

Soon after the establishment of the Presbyterian church, Mr. Akagi, a reporter for *Shinsekai* newspaper, persuaded people to start a Buddhist church. Kameichiro Inoue and others gathered donations, and the Buddhist church was consecrated in 1908 (*Buddhist Churches of America*, 1974, p. 222).

A community was slowly established by the store owners and farmers who settled in the area. Families appeared among them. At the same time, the anti-Japanese movement finally took shape in

Murakami family from Murakami collection,
around 1912
reproduced by Media Services/U.C.S.C.

legal form. In 1907, the Japanese and U.S. governments established a Gentlemen's Agreement. The Japanese government stopped issuing passports to Japanese laborers who wanted to come to the U.S. but allowed wives and children of immigrants already in the U.S. to come. A great number of brides came to join their husbands.

Marriage

Marriage was one of the most important events in an Issei's life but was not easy in this country. Marriage between the Caucasian and Asian races was prohibited by law in California. Those of the Asian race were not eligible to be naturalized at that time. Interracial marriage between Issei and other minorities was rare.[5]

To find suitable marriage partners, the Issei either had to return to Japan or ask their parents or relatives to look for brides. Kumajiro Murakami's parents found a young woman for him in Japan and sent her to the U.S. in 1909. Fuji Murakami, his wife, "grew up in a nearby area" in Yamaguchi Prefecture, though they did not know each other in their childhoods.

For many others, it was not easy to look for their own brides. The picture bride marriage system was practiced to fulfill their requests. It came into the public eye around 1910 and was understood simply to be an exchange of pictures between prospective marriage partners. The *Literary Digest* expressed the general understanding of white people about "picture brides" at that time:

> The Japanese population has been increased a great deal of late by the arrival of brides. Just as soon as a Japanese potato or onion farmer makes a neat sum out of a crop, he writes home for a "picture wife." By return mail, he receives a bundle of photographs from which he selects the most likely-looking ("California Hustling Japanese," 1913).

This may have appeared an odd custom to Americans, but it was the Issei's adaptation of the traditional arranged-marriage system under special circumstances.

Folklorists in Japan consider the village to have been a social and economic unit, where young people found marriage partners for themselves. But marriages at that time in Japan were generally arranged by the parents through a go-between. In some cases, the young couples did not see each other until the day of their marriage ceremony.

The picture bride system was practiced in the same way between couples in Japan and the U.S. The arrangement offered security to a couple, even though they had not met each other. Either their parents arranged their marriage, or they learned about each other from a trusted go-between. Some of them were relatives, and some were from the same area. The Issei exchanged their pictures because they could not see each other, being at opposite sides of the Pacific. It was a valid, legal marriage within the accepted norms of Japanese society, which many young Japanese women accepted without much hesitation. Though such marriages may appear unreasonable, for it meant going abroad to an unknown husband, in reality many Issei women acquiesced to their parents' wishes in good faith.

Early Brides

Masa Kobayashi, who was sent to Tokyo to be educated from the age of 4 to 19, returned to her home in Ibaragi Prefecture after her father gambled away his business. Shuro Kobayashi returned at the same time to see his sick father after being in the U.S. for ten years, from 1907 to 1916. He was already 30 years old. Masa Kobayashi explains how their marriage was arranged:

> My parents and Kobayashi's brother were friends. He asked my father, "Could you give your daughter to my brother?" I was obligated to do so. The old days were like that. After that, we had a premarriage meeting just to see each other's face. But I could not see his face. People these days would laugh at me. I kept looking down. I just looked at his stockings. I thought it would be all right, as the parents said all right.

They had a formal, traditional marriage ceremony at his house in the big room upstairs from the Aburaya dry goods store. Her father asked his son-in-law to let his daughter have time to read books because she loved them so. Shuro Kobayashi returned to the U.S. soon after the marriage, and Masa stayed for a while in Japan to prepare for the trip. She packed only a large *futon* (Japanese bedding), as she was told, and left her *koto* (a stringed instrument) and other memories of her younger days behind. Her mother cried, clinging to a pole in the house when she left. Masa was 23 years old when she came to the U.S.

She arrived in San Francisco in 1916. She rejoined her husband and came to Brawley in the Imperial Valley by train. Although her husband did not lie about their coming life in the U.S. and she had accepted it, life across the Pacific was different from what she had imagined.

> When I saw people with work clothes coming out of their huts, I was surprised. I mean I saw it when I just arrived. The houses looked like huts from a Japanese standard. They were houses made by their own hands.

She describes the house in which she and her husband lived:

> We constructed the framework [of the house] with wood and covered the ceiling with a tent. When it was very hot, we covered the ceiling with grass blinds like the ones people used in tropical areas.

She scooped 40 buckets of water from the pond to the bathtub for the daily water supply and cooked for many men working on the farm. Life was still raw.

> We used many horses. I was scared of them. When I got up in the morning, the horses came out with the men hollering "get up, get up!" I was so scared of them. [Laughter.]

Miteru Mano, a cheerful person, told me how she got married:

> Yes, Papa [her husband] used to work [with her father's cousin]. Since he was such a good boy, a letter came to my father [from his cousin] requesting his girl to be his bride. I did not want to come at all because I did not understand anything. Though my brothers and uncles all were against this marriage, Father gave me away all by himself. He thought this arrangement would not be bad since his cousin chose the groom.

Miteru Mano was born in 1892 in a small farm village in Kumamoto Prefecture. Her family grew everything to eat: rice, wheat, and beans. She did not go to school.

> We could not have a marriage ceremony because the families were so far apart. It was just like here and New York in the United States. I was in Kumamoto, and his family was in Numazu city in Shizuoka

Prefecture, which was almost around Tokyo. I did not know anybody and did not go through a ceremony. We did it only with our family members. It was a picture marriage.

However, formality was valued. Even though her groom was in the U.S. and no close relatives were present at his place in Numazu, she visited his family.

Before I left for the United States, I at least went to Papa's place because I was sent there, as it was the way you were supposed to do things. There was nobody there, not even his brother or sister or anybody, but the people who were adopted into his family were taking care of his house. I at least went there, though. Since I went there, they had a dinner and invited neighbors to celebrate.

In 1913, Miteru Mano was married at the age of 21. She recalls the journey to the United States.

There were many picture brides, about 150 of them. There were many who went to Hawaii. Many picture brides got off in Hawaii, about 100 of them. There were only a few of us who came to the United States.

In San Francisco, she met her husband for the first time.

In San Francisco, he was waiting for me. I had not seen him at all; so I could not tell who he was unless he declared his name. There was somebody who could speak Japanese and English to mediate among the people. There was a man who asked me whether I was Mrs. Mano and did everything for us.

When I asked her if her husband looked like his picture, she just burst out laughing. "I could not understand anything at that time, but I thought he somehow looked like the one in the picture."

Yuki Torigoe speaks energetically with an Okayama dialect. She was 82 years old when I first saw her in 1978, and her cheerful voice has not changed.

Our grandfathers were in the same *kabuuchi*; we were not blood relatives. *Kabuuchi* means relative, well, kinsman. Therefore, we were second or third cousins. Since we were relatives, his father visited us to ask me to marry him. I sent a picture of mine though we were relatives, because he hadn't seen me. I married by picture. And I went through the formal marriage ceremony with our go-between in Japan. I was such a young girl who did not know how to cook. But Papa taught me.

That was how she married him in 1914. He was 30 years old, and she was 18. Both were born in the monkey year of the lunar calendar.

Okayama is famous for *igusa* (rush), and her family used to weave *goza* mats by sending out looms to farmers in the country. She used to soak them in Nikawa glue when she was a child. She finished eight years of education.

When Yuki Torigoe was on the small Japanese boat coming to the United States, she had a pair of *hakama* (pleated skirt that loosely covers the kimono) on, just like a student in a woman's high school at the time, and a modern *soku-hatsu* hairstyle (the hair completely tied up over the head) in the current fashion.

> I could hardly eat since I got seasick. I could only eat some rice soup when it was served. People who did not get seasick could eat all the food served. They even made *sushi* and sold them to us. I remember such a thing.

But a trip to the U.S. was long, lasting about a month, and one of the Japanese women could not manage it.

> There was one who passed away on the boat. I saw the funeral, burying her in the water. She got heavily seasick and died from it. I felt sorry for her, but it could not be helped.

She stayed only one day on Angel Island, the detention center for immigrants, and went through a legal American marriage in San Francisco.

> Next day, we could land at the port. The priests came. They were Japanese priests. A paper just like a diploma was given to us, though I already went through a marriage ceremony in Japan. We all went to the hotel with the priest. There were about ten couples who walked up to the altar.

She was young and excited, looking at everything with curious eyes.

> Everybody was kind to me, giving me advice. Papa was talking to somebody else. Then we took the train. I saw the wagon pulled by horses. I had to climb up to get on the seat of the wagon. There were a lot of them [different things]. When I was in San Francisco, I was at the hotel before I caught a train, where I saw white people kissing at the train station. I had never seen it before. I am telling you the truth.

Her dreams contrasted with actual life in the U.S. Yuki Torigoe was brought to one of the boardinghouses in Watsonville after the train ride from San Francisco.

> I came to the United States happily since I had heard that the United States was a good country. [Laughter.] I had some kind of idea about the United States since I had seen many pictures. I believed that all the houses were grandly made out of blocks and concrete. But when I came to my own house in the United States, it was made out of wood. I thought that could not be. I really thought that way. My papa was staying at Mr. Matano's boardinghouse where he rented one room.

Watsonville still had the rowdy, makeshift vestiges of bachelor life, though changes were slowly taking place by the time Yuki Torigoe arrived.

Abolition of the Picture Bride System

The town of Watsonville slowly changed its appearance. While some stores changed one after another, other stores remained as familiar faces. Among the stores which disappeared were Japanese restaurants where waitresses served traditional Japanese food and *sake*. Those restaurants, named like restaurants in Japan ("Yachiyo-tei," "Asahi-tei"), disappeared as the Issei started to establish their homes.

At the same time, the Japanese in the area, though they were not allowed to become citizens of the U.S., tried to be at least good members of the community and often discussed public morality. In 1907 at meetings of the Japanese Association, Kohachiro Miyazaki (the director of the Japanese Presbyterian church) and Kameichiro Inoue (one of the founders of the local Buddhist church) worked to improve public life with other members.

Their work was continued by such others as the Rev. Suekichi Yoshimura around 1910 (Koga, 1977, p. 184). When a group led by a white man made a public speech about the prohibition of human trade (probably in prostitution houses), irresponsible policemen, and the elimination of gambling houses at the Tōyō Hall, it caused a great disturbance.

> The hall, already full, became packed with the presence of hecklers.

> Some of them from gambling houses and wine stores abused the speaker and threw stones or tiles through windows. Finally, they cut off the electricity line, and the hall became pitch dark (*Shinsekai*, Feb. 25, 1913).

Despite their many attempts, gambling houses and the red light districts remained until World War II.

Around 1920, the anti-Japanese movement escalated, especially among politicians who voted for the Revised Alien Land Law and the abolition of the picture bride system. As the Issei's efforts to establish their economic status and families became more obvious, the improvements the Issei brought to agriculture and the growing number of Japanese brides received closer attention, increasing public criticism. After heated discussion among themselves, the Japanese Association voluntarily asked the Japanese government to stop issuing visas to Japanese women who wanted to come as picture brides. In a difficult situation, the Japanese government decided to protect those Japanese already in the United States and simultaneously to fight the Revised Alien Land Law. But they gave in on the issue of picture brides and halted further immigration from Japan. The Japanese government actually enacted the bill in February 1920. Now single men had to work even harder to save enough money to return to Japan to find a bride.

Late Brides

Eiko Tsuyuki was born in Vancouver in 1903. She still does not know whether her parents registered her birth in Canada or not; therefore, she never bothered to apply for U.S. citizenship.

> My mother came over with my father here when she had me in her stomach. My father was working for a lumber company.

When she was 5 years old, her parents sent her to Japan to be educated. Her mother passed away four years later, and she was raised by her father's twin brother in Yanai in Yamaguchi Prefecture. She had eight years of education, and she learned sewing and manual arts at a special school. She got a chance to go abroad in 1921 when she was 18 years old.

> Kishimura [her first husband] came to Japan to take me to this country.

> Since my [foster] parents had promised, I married into the Kishimura family even before he came back to Japan to see me. Isn't it funny? It was like that in the Meiji era. The parents decided who you were to marry. It was all right with my parents, and I did not know anything about it.

Though it was her foster parents' idea for her to marry Kyōsuke Kishimura, she dreamed of seeing her own father whom she had not seen since she was 5 years old.

> From Yokohama, we arrived at Seattle. My father came over to see me on the ship. I could not land at Seattle; so we saw each other on the boat.

They met only briefly as the boat headed for another port, San Francisco. It was 1921, and many young women were on the boat to the United States.

> I think I enjoyed my honeymoon trip to the United States. My husband did not tell me about the upcoming life in the United States. I was just an innocent girl who listened to and followed her husband as she had her parents before her marriage.

She later told me she had her own dreams and hopes. "As a young girl, I had my own hopes to study English and Western sewing in the United States."

Chizu Matsuoka's name was familiar to me long before I made an appointment to talk to her. She was the only Japanese certified midwife who worked in Watsonville and later in Salinas from 1925 until the start of World War II. She does not remember how many babies she helped deliver.

Chizu Matsuoka was born near a coal mining village in Fukuoka Prefecture, near Beppu spa in Oita Prefecture. Her village was located near the end of the railroad line. People thought that children did not need an education to grow up in a farming village. But she wanted to be a woman with a profession. She was the only woman from the village to go to *kōtoka* (two years of post elementary school education) school, and she even went to Hakata City (presently Fukuoka City) to study when she was still a young girl.

When she returned to the village, she worked for an office which treated injured miners. After receiving notice that she passed the test to become a nurse in 1924, her future husband, Tonai Matsuoka, a Yobiyose Issei, came back from the U.S. His father and her father were cousins. Her marriage to Tonai was arranged by their parents before he came back. Besides her parents' decision, she had her own individual dream, and she wanted to go to the U.S.

> I had a dream in this country. There were many people who wanted to come to the U.S. at that time. I did not know any negative part of this country. My husband's father used to visit us, and I knew he had a comfortable life after returning from the U.S.

She went through a formal marriage ceremony in Japan. They were in a hurry to go to the U.S. as he was at a draftable age. She landed in Seattle in early 1924. Everything was rushed and done in a flurry.

> I was soon taken into the hotel, where everybody took care of me and helped me to dress up in the Western clothes. I was totally new to everything and had to give free rein to the salesperson, as I did not know the value of any of those items. I had a corset and was taken to a hairdresser, too. I stayed in Seattle only one day for those preparations.

By the time she came to Watsonville, the town was well established, and her husband had the Yokohama laundry store in the middle of the Japanese residential area. She was busy helping her husband, but she was an energetic woman.

> There was a night school here, but I had to work. I could go to night school for one or two years, but it was not enough because I had forgotten all that I had learned. I did not have any chance to practice English here. Everything was transacted in Japanese among Japanese here.

She did not give up. She continued studying and applied for the test to be a midwife in this country.

> It was just amazing that I could pass the test to be a midwife in this country, considering the difficult problems I had to solve in the test which were so different from the ones given in Japan. There were a lot more problems to solve in this country. . . . I was the only Japanese among all the examinees. Each answering sheet was bound like a book, and the next page to my answer in Japanese was blank, so somebody could translate my answer into English. There were four subjects: anatomy, sterilization and disinfection, pediatrics, and general

midwife practice. I had to answer all those questions within two and a half hours. I was really exhausted when I finished it. I was anxious and worried a lot about the announcement for a week, and I was happy that I passed.

Yobiyose

As many young brides were taking a long boat ride to this country, many other young men and women were called by their parents to help them in the U.S. Yobiyose are the pioneer Issei's children who were left in Japan until they became old enough to work. Many Yobiyose were still in their teens when they were called to the U.S. Although it was their parents' decision, these Yobiyose had their own dreams to fulfill in this country.

By the time many Yobiyose came to Watsonville, the Japanese community had been established, and many Issei had families and businesses. On the other hand, some Issei were still transient farm workers who stayed in the boardinghouses. Around 1920, the labor clubs changed into communal residences for transient people, and the young Yobiyose who came to the U.S. at this time knew little of the original purpose of the clubs. Toshi Murata, a young Yobiyose at the time, explains:

> Most of the people, who came to town to stay in the boardinghouses, stayed here only for one season and left after the season was over.

If Issei farmers needed help on their farms, the boardinghouses could supply young men. The number of boardinghouses slowly declined, but young Yobiyose men still occupied them. Young Yobiyose worked on the farms and in packing houses whenever jobs were available.

Masao Kimoto was only 14 years old when he was called to the U.S. by his parents in 1914. He was a junior high school student in Japan. On the boat journey, he slept in a big room with other people until he landed at Tacoma. "My father came to pick me up. I love boats, so I came down by boat here. The boat landed in San Francisco."

His mother had been called to the U.S. a few years before him. She could not come to meet him, as she had just given birth to his younger brother, Naoyuki. His mother had a baby girl, Chiyoko, three years later. Masao was only able to attend school the first year after he arrived in the United States. Soon he began to help his parents on their strawberry farm and worked for the C. C. Morse Seeds Company (now Ferry Morse) during the summer, staying at the boardinghouse in town.

At the age of 22, Masao Kimoto saved up $1,000 working overtime for three months in the apple packing house; he used this money to return to Japan to find a bride. He not only paid for the transportation but for his bride's preparation as well. Before he left, he had a definite person in mind.

> She was a sister of the wife of an acquaintance whom I happened to get to know in San Francisco. He offered me his wife's sister. And I told him I would marry her if both of us liked each other after I met her in Japan.

He liked her and went through a formal marriage in Japan.

> She was teaching sewing after she graduated from women's high school because she was good at sewing. Then she became a teacher at a kindergarten. She did not know I was coming. Though I asked the go-between to arrange it formally, I personally discussed it with her sister. I told her that I came to propose to her sister. Then I proposed to her. I told her that she had to work on the farm, as I was a farmer. She did not say "no." She had never worked on a farm because she was a daughter of a contractor.

On the boat to the U.S., he did not want to sleep in the big room where people were squeezed together. Instead, he took a small room, so he could have more privacy with his bride. He said, "I had a happy bon voyage."

Toshi Murata came to Castroville in 1921 when he was 14 years old. "As a child, I thought the U.S. was a good place." Castroville was windy and lonesome. He could not find anybody his own age; other Issei were all much older, and the Nisei were too young for him to play with. His older brother, who had come to the U.S. four years earlier, took him to public school where he learned English for one year and a half.

In 1923, when his father returned to Japan, his older brother decided to stay in the U.S. to start his own business, leasing some land. So Toshi Murata also stayed. At 17, he worked at his sister's farm in Watsonville for one year then became a hired farm worker in the Watsonville area from 1924 to 1927.

Around 1923, the Issei were keenly aware of the coming immigration laws and urged young Yobiyose to return home as soon as possible to find a bride. But Toshi was too young to marry and settle down. "I loved to travel." He was young and healthy and traveled around California with "just simple stuff" without blankets and cooking utensils.

> Mostly, I went around by myself. Since that time, I have loved to travel. . . . At first, I bought a motorcycle and drove it to many places. I liked it.

In town, he stayed in a boardinghouse, but when working on a farm, he stayed at the camp house provided by the employer. He talks about an apricot packing shed where he worked: "That was near Tracy. This was also owned by a white person, and a Japanese boss made a contract and managed the camp." Outside Watsonville, Japanese bosses mediated between laborers and employers, "took care of the boarding, and provided the meals with which they could make some profit." Married couples who traveled around were usually given a separate space. When he was moving around, he could have a bed with a white sheet in the sleeping corner of the camp house.

Single men were free from responsibility. Toshi Murata worked for one season then traveled with that money for two or three months until the money ran out. "I did not care at all. I always thought I had tomorrow. That was the way I was."

Uneasy Settlement

Growth of a Community

> There is nothing more ridiculous than hearing what Senator Phelan has to say about keeping Japanese on low wages and separating Japanese from the land; so they will not threaten the standard of living of the average American. On the contrary, this would be the biggest threat of danger to California state, to make Japanese work under such low wages like the ones they received in their own country, and force Japanese to endure an unsanitary environment and poor food. Japanese are also human beings (*Shinsekai*, April 29, 1919).

The times were changing. Modernization brought greater comfort to the lives of the Issei. Many white immigrants were finding their way to settlement in the U.S. Some of them obtained citizenship and eventually owned their own small farms. Among the 4,809 foreign-born whites 21 years old and over, 2,351 were already naturalized and 314 applied for their first papers in Santa Cruz County, according to the 1920 census. In town, people abandoned their wood stoves for gas stoves. Yuki Torigoe, who lived in the middle of town, remembers:

> I had a gas stove when I had the second girl. I think I had a gas stove in my house around 1916. I dried and warmed diapers by charcoal in winter [for the first girl].

In the country, the *kudo* was no longer used by the time World War I was over. Ichiro Yamaguchi, born in Pajaro in 1908, remembers:

> *Kudo* had gone. The last time I can remember using it was before the war, World War I. They used regular natural gas, yeah, propane gas just before the war. Those came in, and they were using a wood stove for the heat during the winter. But most of them had a regular wood cooking range. If it got hot, they used kerosene.

Transportation was also modernized. Although most people rode bicycles to work, the number of car owners slowly increased. Yuki Torigoe remembers:

> In 1915, by the time my big girl was born, there were quite a number of people who went to San Francisco to buy old cars.

Tokushige Kizuka, a young Yobiyose Issei, was called to the U.S. in 1917 at the age of 16. He remembers that his parents, who bought a car and truck in 1918, were among the first car owners in the area. By 1924, many Issei had their own cars. The number of boardinghouses decreased as men married and lived in their own houses, and people outside of Watsonville could visit town on a one-day trip.

Entertainment was found not only in billiard parlors or Shanghai Banks but also in the Tōyō Hall which provided Japanese movies and concerts ranging from amateur *kabuki* to *gidayu* (a kind of Jōruri, a singing story-telling with *samisen*). Women *gidayu* singers were treated like idols by young men in Japan from the Meiji era to the Taisho era. There was entertainment for everybody. One Issei remembers that Hibari Misora, a popular Japanese singer, came to Tōyō Hall as a young girl, and a famous classical singer also came. As the number of families increased, so did the number of family-oriented community events. Once a year, the Japanese Association planned picnics at the beach. Both Buddhist and Christian churches organized events for their members.

At this time, the Japanese community developed greatly. The Issei established families and invested more in independent businesses. The Japanese population in Santa Cruz County increased from 689 in 1910 to 1,019 in 1920. However, the anti-Japanese sentiment was growing, and the state openly tried to repress Japanese immigrants through laws and regulations. The Issei's efforts to rise above the harsh life at the bottom of the California farming system were vigorous and had almost succeeded. But now the Issei had to surpass more obstacles set up by the state. The narrow path that had been opened to them now became even narrower.

Neighbors from Torigoe collection, 1917
reproduced by Media Services/U.C.S.C.

The Alien Land Law of 1913

The first Alien Land Law was issued in 1913 in California. This law prohibited Japanese from owning or leasing land for more than three years. The land in Pajaro Valley was rich and fertile and there was a great demand for it. Only a few Japanese succeeded in owning land before the land law was issued. Strawberries, the major crop of the Issei in this area, were not productive during the first years and could not bring enough profit within the limit of three years. Therefore, Japanese farmers usually stayed on the same land for four to six years.[6]

The *Japanese American Yearbook* provides detailed documents on Japanese farmers. In the Watsonville area, the average size of farms decreased from 15.8 acres in 1910, to 14.9 acres in 1911, and to 12.7 acres in 1914; the number of farms increased from 155 to 165 and to 190 during the same years.

Further examination shows surprising changes among Issei farmers from 1910 to 1914 in Watsonville. The number of independent farmers who either owned, leased, or shared farms decreased substantially from 95 in 1910, to 107 in 1911, to 45 in 1914. More surprisingly, farms under three acres numbered only 4 in 1910 but 44 in 1914. All these farms in 1914 were under contract and cultivated strawberries. The number of contract farmers increased from 6 two-man partners and 54 individuals in 1910, to 2 two-man partners and 56 individuals in 1911, and to 2 three-man partners, 12 two-man partners, and 131 individuals in 1914. We cannot determine the accuracy of the statistics in the *Yearbook*, but these numbers show the considerable impact of the Alien Land Law. Now that the Issei's business investments were discouraged, they were forced to remain small-scale farmers.

Strawberries were one of the most intensive crops. Farmers could make some profit on even two acres of land. The Issei and strawberries in the Pajaro Valley became inseparable. In 1914, the Watsonville area, including the San Juan and Hollister areas, had 1,500 acres of strawberries, and all were cultivated by the Japanese, although 60 percent were managed by white farmers (*Japanese American Yearbook*, 1914, pp. 117-18).

Newborn American Citizens

In agricultural areas, the landowners, who were usually white, provided housing for the tenant farmers. Many Japanese sharecroppers lived in shacks with walls constructed of unpolished boards. The crevices in these walls were covered from the inside by newspapers to prevent cold drafts. For easy construction, all the houses were the same size with the same unpolished boards for walls, roof, floor, and ceiling if there was any.

Japanese sharecroppers who had a contract with the same landowner or boss lived close by each other and created their own colonies. These camps were much inferior to the camps of white farmers, and newlywed wives of Issei farmers must have suffered great disappointment, though most were able to cope and accept the conditions of their lives.

Eiko Tsuyuki, the wife of a half-share tenant farmer, who came as a bride in 1921, remembers her shelter:

> Strawberries needed new land every three years at that time, and we had to move our houses with us as we changed location. Therefore, the house we lived in was not at all [a] house, but a temporary shelter. I could see the sun through the boards in the ceiling, and the wind would come through between the boards on the wall. Later, we received wallpaper to close the cracks between the boards.

These houses lasted for a long time and were used by the Pilipino sharecroppers even many years after World War II.

Issei women could not afford to stay home but had to work in the fields next to their husbands. In growing strawberries, there was always something to do, and weeding was an important task for the women. Some women learned to be strong. Masao Kimoto remembers his mother negotiating with men to work for them:

> When we were growing strawberries, my mother went there [to boardinghouses] to get men, telling them that she would cook for them.

Issei women worked outside during the day and took care of the household matters at night. They had to take care of the children, prepare all the meals, and they did the wash by hand. There was always something to do. Eiko Tsuyuki says:

I believed that was my duty as a woman. I never thought of it as a burden.

Even though these temporary shelters were in terrible condition, it was much better to have their own home than to migrate with their husbands following the crops. But the demanding lifestyle made it difficult for them to begin families.

> Nowadays, about 400 of our men are already married in the Watsonville area, and approximately 80 babies were born a year. Nevertheless, it is a problem we could not overlook for the future development of our immigrant people that ⅓ of those babies died. The reason why the infant death rate is so high is that they could not pay attention to hygienic treatment after birth, and it is said that the strawberry business, that our people specialized in, forced pregnant women into that condition (*Nichibei Shimbun*, Nov. 9, 1913).

According to this article, Japanese mothers gave their babies cows' milk or powdered milk instead of breast-feeding and fed them irregularly. The writer of the article warned women who were eight or nine months pregnant not to work long hours in the cold and wet strawberry fields. One white doctor warned that

Deserted sharecropper's house at Shikuma residence
photo by Mona Nagai, 1983

many one-week or nine-day-old Japanese died of pneumonia, caused by carelessness after bathing them (*Nichibei Shimbun*, Jan. 22, 1915). Cold houses and severe conditions were to blame, not the mothers.

Many women suffered from women's diseases, especially after delivery, since they were not informed about proper hygiene. In the early days, some women had no medical care after they delivered a child. In some cases, their husbands helped them, and occasionally their mother-in-law helped. Many people did not bother to call a doctor or certified midwife, whose reports sent to the county records office would prove their children's American citizenship. Even Chizu Matsuoka, who was a midwife in Watsonville from 1925, remembers:

> Women always had to work on the farm. Mother took a small baby on her back to the field and left the baby in a box next to the field while she was working. Small children were running around in the strawberry field. White people used to say that Japanese reared children in the field as if they reared chickens.

She also said that children were born every two years in each family. Those children became the major reason why the Issei could escape the situation forced on them by the Alien Land Law of 1913.

Elaine Murakami (1974) studied the effects of the Alien Land Law on Issei farmers in Watsonville. She found that birth registrations were greater than the number of actual births around 1914. Issei parents who had neglected their children's birth reports now registered them late, in order to use their names for farm contracts. Their children, unlike themselves, were Americans by birth with full citizens' rights.

Mack Shikuma said, "I had trouble about my birthdate." He reasons that his parents did not know how to register his birth when he was born in 1907. Another reason could be that his parents were not sure at that time if they were going to remain in the U.S. or return to Japan.

According to Murakami, Japanese-registered childbirth between 1887 and 1920 was the highest in 1914 with 126, and in 1915 with 127. This was far above the actual number of births, 94 in 1914 and 102 in 1915. Finally, neglected childbirth reports in the county records office almost caught up to actual births.

Miteru Mano, who came to the United States in 1913 and worked at the vegetable market with her husband in San Luis Obispo, had ten children, five boys and five girls.

> I have never been in a hospital since I was born. I had so many children, but I had all of them at home and did everything at home. But I had to call a doctor for their birth certificates to get American citizenship.

Now the Issei could invest in their farm businesses by becoming guardians for their children who had full citizenship rights. They could then sign a contract for the farm. The *Japanese American Yearbook* indicated that independent Issei farmers (excluding contract leases) increased: 45 out of 190 in 1914 and 125 out of 161 in 1915.

Recovery

The First World War (1914-18) dramatically improved the economy in the United States. Several Issei mentioned Yugoslavian people becoming the owners of small apple orchards and packing houses by 1920. The upswing helped the Issei farmers, and it did not take long for some Issei to get their independent businesses back.

During the war, navy beans were in high demand, and in 1918, farmers in the Watsonville area, including San Juan and Hollister, cultivated 2,497 acres of beans and 443 acres of strawberries. Many contract farmers moved up to manage leased land, and the *Japanese American Yearbook* (1918) reported one owner and 172 cash leases. These included 190 men, 162 women, and 312 children. On the other hand, it also reported that hired farm laborers numbered 166 men, 70 women, and 70 children.

The improved economy made some Issei magnanimous, and they invested in large businesses. It was as if they were waging a gamble which could take them all the way to victory or defeat. Kumajiro Murakami explains how Y. Kōsansha became bankrupt:

> I was working for the Kōsansha company until it went bankrupt around 1915 when the great exposition took place in San Francisco. We counted on the fair in San Francisco and expanded our business too

much. People did not buy strawberries more just because of the fair. We had to start farming individually then.

Mr. Murakami is an optimistic man who loves horses and singing. He was often asked by strawberry farmers to repair irrigation ditches and would work with his horse, singing loud *minyo* (Japanese classical folk songs) or *shigin* (Chinese classical poems) in the field. Around 1920, he managed 20 acres of strawberry fields on leased land.

Tokubei Kizuka, the father of Tokushige Kizuka, had lived in the Watsonville area since 1907. In 1914, as one of a few Issei growers who raised another crop besides strawberries, he and his partner, Matazaburo Mine, signed a contract for over 40 acres of apple orchards. By 1920, when he returned to Japan as a success, he and his partner were leasing 110 acres of apple orchards.

In Lompoc, Harry K. Sakata made a very good profit from navy beans with the help of a Jewish broker who gave him his English first name. Frank Sakata, his son, recalls one day when a white man came to his father's farm:

> Good land was scarce in Lompoc. In 1917, an American man visited him, offering to buy his land for $1,000 per acre. He thought the American man was joking. He did not want to sell the land and simply told him to bring cash if he wanted to buy the land. The next day, the man brought $20,000 in cash. His two cousins wanted to sell the land so that they could go back home to Japan. He had to give in.

With this large sum of money in hand, his two partners eventually left for Japan with proof that they had "raised banners" (a symbol of success) in the U.S. But Harry K. Sakata, who was determined to stay here from the beginning, had to look for new land. He even went to Mexico. On his way back from Mexico, he visited a friend in Watsonville and decided to stay.

When Mr. Sakata came to Watsonville at the end of 1917, strawberry cultivation in the Valley was very productive, and Issei growers were major contributors to its success. The growers, however, had to rely upon commissioners for marketing, which unfortunately was undependable.

According to Wilhelm and Sagen (op. cit., p. 207), the Central California Berry Growers Association was organized on April 9, 1917, under the leadership of Richard F. Driscoll. But, interestingly, the *Japanese American Yearbook* (1918) states that the association was originally advocated by Issei who were members of the Kashu Chūō Nōkai (California Central Farmers Association). Considering the relationship between white growers and Issei growers at that time, it is most likely that Issei growers contributed greatly to the start of the California Berry Growers Association and succeeded in involving white growers to work together, rather than the other way around.

Many local berry growers signed a contract with the association, which became their agent and took care of the consigning, marketing, processing, and preserving of berries. The association even had its own freezing plant.

> [T]he Association, in 1921, had built its first freezing plant for strawberries at the Security Warehouse Company of San Jose (ibid., p. 210).

With the help of the association, growers could better cope with the fluctuating trends of marketing. A number of growers thus secured marketing for their crops. In 1922, the Central California Berry Growers Association adoped "Naturipe" as a trademark and eventually took the name "Naturipe Berry Growers" in 1958. The association's bylaws indicated the importance of the Issei in strawberry cultivation. According to Wilhelm and Sagen, "Bylaws of the association adopted on that day called for ten directors, five to be Japanese-American and five to be Caucasian strawberry growers" (ibid., p. 207).

Unosuke Shikuma was one of the most well known and one of the largest Japanese strawberry farmers. Mack Shikuma, his son, said that his father made good money from World War I. Unosuke, a member of J. S. Kōsansha, started to invest in separate businesses during the war. He invested in the Sunset Berry Farm (32.5 acres) as a partner for two years from 1916, and the K. U. S. ("K" stands for Kōtaro, his uncle, "U" for Unosuke, and "S" for Shikuma) Strawberry Farm (60 acres) for four years from 1917 (*Japanese Who's Who in America*, 1922, p. 550).

Besides managing the leased farm in Watsonville, Unosuke Shikuma increased the scale of his business to over 200 acres with the Oak Grove Berry Farm in Salinas in Monterey County. He shared this enterprise with partners Matazuchi and Heizuchi Yamamoto (brothers), and Henry A. Hyde and Orrin O. Eaton, local

Caucasians. Mr. Shikuma had known these partners long before this time, and it was beneficial for the white businessmen to incorporate with the big Japanese berry growers who could bring a large number of sharecroppers to work in the fields. Also, this partnership was strengthened by their relationship to the Central California Berry Growers Association. Unosuke Shikuma was an active member of this association.

According to Wilhelm and Sagen (op. cit., p. 204), the Oak Grove Berry Farm was one of the biggest strawberry farms in the world.

> The ranch was planted about 1920 and was an outstanding example of water engineering, land leveling, scientific culture, shipping, and management of a few hundred workers. For several years it was the largest and most productive strawberry field in the world, and had an output of over three million baskets per year.

However, the farming methods used in the fields of the Oak Grove Berry Farm were no different from any other strawberry farms in that area, even though the preparation was done with modern science, e.g., water engineering and land leveling. This huge land was divided into smaller sections and Japanese sharecroppers raised strawberries in these sections. The Yamamoto brothers became foremen overseeing the farmers.

World War I brought great success to a small number of the Issei, and those who gained a degree of independence caught a few drops of the then good economy. However, the majority of the Issei needed their own people's help. Tokushige Kizuka, who was called from Japan by his father in 1917, seems to explain how dependent they were upon each other at that time. He said:

> At that time, people did not have capital, and the banks would not loan the money to them like in these days. They borrowed the capital from a person who had it and something like that. In case somebody unfortunately had to face bankruptcy, all the other people had to suffer difficulties from it in the old days.

Revised Alien Land Law of 1920, 1923

When many soldiers came back home after fighting in World War

1, one Issei veteran in Watsonville applied for citizenship. Although aliens were permitted to apply for citizenship if they served voluntarily in the armed forces, Frank S. Becker, United States naturalization examiner, "refused flatly... to admit the Watsonville Japanese" (*Watsonville Evening Pajaronian*, March 27, 1919). Nevertheless the news article sided with the Issei.

> If this man came out for the old flag and bared his breast for this country in the big War... he is entitled to citizenship whether he be Japanese, Chinese, or any other nationality.

But the situation for the Issei was acutely aggravating. Anti-Japanese feelings had broken through the friendly surface of the community. During this time, the Issei tried to counteract these feelings by supporting the community and buying large numbers of war bonds. The *Japanese Farmers in California*, published in 1918, noted (p. 30) that the Watsonville Japanese Association alone purchased $8,150 in Liberty Bonds. The *Watsonville Evening Pajaronian* (April 28, 1919) indicated that K. Kamigaki, who owned a grocery store on Main Street, had taken out a $500 Victory Loan; K. Yoshida, $250; Nippon Shōkai, $200; K. Kitaji and N. Takashugi, $100; as well as others.

The anti-Japanese movement was manifested in the local *Watsonville Evening Pajaronian*, which constantly brought up news about the Japanese. An April 4, 1919, article entitled "Japanese Question to [President] Wilson" said:

> Request has been made to the senate of California for permission to introduce two bills. One absolutely forbids Japanese to lease agricultural lands in this state. The other is directed against the so-called Japanese picture-bride marriages.

As indicated earlier, the Japanese government had decided to stop issuing passports to picture brides, hoping to stop enactment of an alien land law. After that, a Revised Alien Land Law was repeatedly discussed by politicians. In 1920, Sen. J. M. Inmann, president of the California Oriental Exclusion League, actively tried to place the anti-Japanese land measure on the November ballot.

> [T]he initiative petition [was] signed by thousands of voters, and another is that sentiment in California is almost unanimous

for legislation that will stop the Japanese and other orientals from getting possession of the best agricultural lands in the States (*Watsonville Evening Pajaronian*, Aug. 23, 1920).

Politicians proclaimed support for the Revised Alien Land Law in campaigns for the U.S. Senate and Presidency. Samuel M. Shortridge, one of the candidates for U.S. Senator, made a speech in downtown Watsonville on the afternoon of Oct. 28, 1920 (announced in the *Watsonville Evening Pajaronian*, Oct. 23, 1920). His message to the people of Watsonville and Pajaro Valley was:

America First; the Japanese must not come; and
Protect American Products for the American Market.

The campaign was vicious. During the general election on Nov. 2, 1920, Initiative One passed in Santa Cruz County by a great majority. The Revised Alien Land Law was put into effect on Dec. 9, 1920. The Issei, who were not eligible for citizenship, could no longer legally lease land and could not become guardians of Nisei children unless the court gave special permission. Furthermore, in order to apply the law specifically to Japanese, it was revised again in 1923.

The Issei in Watsonville area felt strongly about the Revised Alien Land Law of 1920 and 1923. "It is indeed miserable to see our people just become a loss under the land law" (*Shinsekai*, Nov. 3, 1924). All the Nisei children were still very young, and none of those who were born in Santa Cruz County had reached legal age by 1920. The Issei in Pajaro Valley had to think of some way to continue their businesses.

Harry K. Sakata established a family corporation in Watsonville with the help of an attorney. His son, Frank Sakata, explains:

In 1917 when he could not buy land, he received advice from a lawyer, attorney Guy C. Calden, who specialized in helping the Japanese. The lawyer advised him that the way to circumvent the law was to form a corporation of which the majority of stockholders had to be Americans. Therefore, he and his wife held 48% of the stock, and his children, including myself, who had American citizenship, held 52% of the stock. Since we were under age, over half of the stock had to be held in trust by the lawyer until we were of age.

With the help of G. C. Calden, the Sakatas organized the L&W

Company and purchased 60 acres of land. Mr. Sakata cultivated various vegetables for a couple of years then decided to specialize in lettuce.

Tokushige Kizuka stresses that he had known a white lawyer from his parents' generation who was involved in his land purchase in 1926 or 1927:

> [Mr. Hogan] offered me his land for sale.... At that time, the Alien Land Law had already been issued, and I could not buy the land. Mr. Hogan told me, "it doesn't matter about the Alien Land Law, don't worry about it." He told me to discuss it with Mr. Son. So, I went to see Mr. Son, explained what Mr. Hogan said to me, and asked if I could buy the land. Then he said, "It would be all right, and I will prepare the papers for it." He told me that I do not have to worry about anything as long as he is alive. I bought the land without worrying about it.

But most Issei had difficulty finding ways to get around the new Alien Land Law. Several Issei commented that the relationship between white and Issei in the area had been "pretty good." It could have been because most Issei never became totally independent, but remained hired laborers, sharecroppers, or at best, lessees of farm land. In other words, they were always dependent on the white farmers and never became real competitors.

According to the 1920 U.S. census reports, 74.1 percent of the farms in Santa Cruz County were operated by owners, and the average farm was 82.3 acres. At the same time, substantial numbers of Issei remained in Pajaro Valley for many years. Farm contracts between local farmers and the Issei could develop on a personal level.

Until their children reached legal age, the Issei had to rely on white farmers for farm contracts. Mack Shikuma remembers:

> At first he [his father] had a problem until I got to be of age. Then I remember the landowner had a good understanding with my father, so that everything goes to the landowner, and the landowner paid him later.

The Revised Alien Land Law of 1920 and 1923 could not destroy those verbal agreements. Also, it could not stop the Issei from leasing land in an American citizen's name. But these verbal agreements made without legal contracts were not necessarily carried through and often caused trouble among those involved.

Toshi Murata, who came to Castroville in 1921 and moved to Watsonville at the end of 1923, commented on the Issei farmers in Pajaro Valley. "It was as if the Japanese were given special permission to grow strawberries." In reality, it was impossible to wipe the Issei out of the fields in the Valley. The Issei were very important workers, and they had been raising one of the most important crops, strawberries, almost exclusively. Many Issei were sharecroppers in the fields and were not bothered by the Alien Land Law. Toshi Murata explains:

> Men who managed on a large scale in the leased land borrowed their name from somebody else, but the small farmers who had share arrangements had verbal contracts with white landowners, which were not made public.

This was the situation of many Issei farmers. Several Issei indicated that when they tried to request compensation from the government after the bitter camp experience of World War II, many of them did not have written documents to indicate their actual losses.

The Issei's dependence on local farmers grew stronger. Even though they had a strong desire to escape that dependence, they had to wait patiently until their children reached legal age. After the Alien Land Laws, the Issei were forced to remain sharecroppers and they had to live in their own communities.

So Far Away from the Mother Country

Even with the Alien Land Law, the anti-Japanese movement did not die away. Next came proposals to reduce the number of Japanese by prohibiting immigration to the U.S. The anti-Japanese emotion was strongly expressed, and the children of local Japanese in Watsonville faced discrimination.

> Until last spring, Japanese children could also attend [the public school] just like white children without any apprehension. But perhaps inflamed by anti-Japanese emotions, the school district tried to segregate Japanese children into one building, a school in name only, since it was really a shack on Amesti Road, using the pretext of classroom shortage and lack of teachers (*Shinsekai*, Feb. 1, 1924).

Although the time and place of relocation differ in reports by *Shinsekai* (Feb. 1, 1924) and *Nichibei Shimbun* (Jan. 31, 1924), both noted that Kōtaro Shikuma, Kouemon Fujita and a few local Caucasians fought the school segregation problem. Because of their efforts, by the end of January 1924, Japanese American children were reinstated and allowed to attend the public school with other local children.

The Japanese Association was busy too with issues of immigration law besides its regular tasks, such as exemption from Japanese military service for young Issei. They warned the public of the upcoming immigration law. A number of Yobiyose went hastily back to Japan to get brides, and some Issei who had left families in Japan had to decide whether they should call them to the U.S. or return home.

The Immigration Law of 1924 passed the House by 308 to 62, and the Senate by 69 to 9. The President signed it on May 26, 1924. The bill set limits on immigration, giving a quota to each nationality based on the eleventh census of 1890, when the Japanese population in the U.S. was relatively small. It practically prohibited Japanese immigration to the U.S.

Around 1913, some people were still strongly tied to Japan and thought of going back home. The *Nichibei Shimbun* (May 9, 1913) indicated:

> The Watsonville Japanese Association asked the health department in this town about the reports of the childbirths from our own people, which were all together 150 for eight years from 1905 to the present [1913]. They estimated 250 birth reports exclusive to Japan.

There were more birth reports sent to Japan than recorded in the United States before the Alien Land Law was issued in 1913. Two hundred and fifty Nisei had Japanese citizenship. A majority of 150 Nisei who registered their births in this country obtained dual citizenship.

The height of immigration was in 1907, and by the 1920s, many Issei had already established their families in the Pajaro Valley. It was becoming their home. The Issei were concerned about their growing children. *Shinsekai* listed many cases of Issei withdrawing their children's Japanese citizenship in 1924. Masa Kobayashi withdrew her children's Japanese citizenship because she hoped the children would become "real Americans."

Tominosuke Umino, who bought 35 acres of land in 1920, withdrew the Japanese nationality of his first son, Sadao, in April 1924. Others, including Yoshitaro Ono, Unosuke Shikuma, Shuro Kobayashi and Harry K. Sakata, withdrew their children's Japanese citizenship. Many of their children were still young. Frank Sakata, Harry's oldest son, was 13 years old, and Tetsuya, Shuro Kobayashi's oldest son, was only 6 years old in 1924. Unosuke Shikuma's third son, Hiroshi, was only 4 years old at this time. Some Issei did not bother to register the births of their newborn children in Japan.

> There were some among our people in the Watsonville area who registered their children's birth in the U.S. but did not bother to do so for Japanese citizenship. Recently, there were ten of these cases (*Shinsekai*, May 29, 1924).

The year 1924 brought another big change to the Japanese community. Prohibition, which had been discussed throughout the previous year, was enacted in 1920 and lasted until 1933. The Kitaji *sake* store in the Japanese residential area was changed to a soft drink parlor. County officers raided gambling houses in Chinatown in Monterey and Salinas. The *Watsonville Evening Pajaronian* (Nov. 26, 1923) announced: "Gambling in Watsonville put under ban by new order of Chief of Police."

There was a movement among local white residents in Watsonville to clean up Chinatown so that a road could be built straight from Main Street in Watsonville to San Juan Road in Pajaro, instead of bending around Chinatown. The *Evening Pajaronian* (March 1, 1924) ran a front-page article, "$850.00 Fire Sweeps Local Chinatown."

> The heart of Chinatown, just across the Pajaro River from Watsonville, was entirely wiped out early this morning by a fire which raged over three blocks, and tore a great, black, smoldering pathway from the San Juan Road to the river "Jungles," south and west.

Since the article stated there were three simultaneous fires started in Chinatown, there was good reason to suspect arson.

After the fire, the Chinese businessmen who lost their houses moved into the Japanese residential area, and it became a racially mixed neighborhood. Slowly, young Pilipinos started to move in, too. Nisei children grew up with Chinese American children, and one Chinese American who spent 11 years of his boyhood in Watsonville remembered it as "an example of early harmonious diversity." The area, however, was predominantly Asian American.

Nothing Left in My Hands

Difficult times continued after the passage of the 1913, 1920 and 1923 Alien Land Laws and the 1924 Immigration Law. The Great Depression made it even harder for the Issei to improve their situation. Masakazu Iwata explained the impact on Issei farmers in *Agricultural History* (1962, pp. 31-32):

> The agricultural depression of that decade may have been a weighty factor in forcing some of the Issei to leave farming, but the legal barriers, the land law and the Oriental exclusion act of 1924, did much to discourage the Japanese from entering farming or expanding their operations.

The above is very true; on the other hand, where else could the Issei go and what could they do? Iwata explores this question:

> The Japanese, in contrast to the Chinese, were unable to enter manufacturing establishments in the cities, such as the cigar, shoe, and clothing factories, partly as a result of the earlier agitation against the Chinese who were employed in those industries and because there was an adequate supply of European immigrant labor for these jobs (ibid., p. 27).

After all, agriculture was the least preferred occupation, even among immigrants. Agricultural laborers, who were not unionized, were paid less than industrial laborers (Fisher, op. cit., p. 12). Many Issei in cities managed small family businesses. But it was not easy for rural Issei to move into cities, look for jobs, and start new businesses.

Harry K. Sakata, one of the most successful businessmen, did not suffer from the Depression but made some profit. Frank Sakata explains:

> Though other crops suffered during the Depression, he earned enough

money to buy more land, receiving a good price on lettuce in summer all over the United States. As one result, he purchased Gilroy Hot Springs in 1932.

But generally, businesses could not sustain themselves as the economy tumbled. Oak Grove Berry Farm, the world's biggest strawberry farm, went bankrupt. Unosuke Shikuma kept an independent business in Watsonville with the help of his sons. His oldest son, Mack Shikuma, was 20 years old in 1927. They bought 20 acres of land in Corralitos in 1927 and hired sharecroppers to cultivate it.

This was no longer the time to gamble in business. Masao Kimoto started his own business at the age of 25 in 1925, three years after his marriage. He explains that, "You must keep your business going; you must always have goods for sale." He gives an example:

> I grew strawberries before lettuce. If I grew three acres of strawberries the first year, then I planted another three acres of strawberries next year. I always had enough. I may either have ten acres or a few acres, but I always have to have goods to sell. If you have old plants, productivity will decline, and the quality will become third class. If you want to have good quality, you have to have young plants. Dealers want to get the best, freshest ones, and would buy other second, third class ones with a fair price along with the first class.

Mr. Kimoto became a careful, steady business person. He learned from experience when he first planted lettuce.

> Well, at first I made a mistake. The reason I made a mistake was that I planted lettuce all at once, being ignorant. I was going to sell them just like in gambling, the whole 25 acres of lettuce at once. Unable to negotiate the price of lettuce, I could not sell them.

In hopes of catching the last wave of an earlier prosperous economy, Kimoto took a chance and lost. Even a careful farmer could not survive hard times, and he later became a transient hired farmworker.

Shuro Kobayashi and his wife, Masa, moved to Watsonville sometime after 1920. They had had a hard time in the Imperial Valley. Mr. Kobayashi explains:

> The reason we left Imperial Valley, where we lived for a long time, was that all the profit went to the commissioners, regardless how

many years we worked. At the end of every year, we would be in the red and were forced to borrow money for the next year. There was no end. Among the people who could not stand it, some left for the city, and others left for other areas.

Life in Watsonville was not much different from that in the Imperial Valley. Kobayashi and his wife became sharecroppers on the berry farms but later raised berries on leased land, using Mack Shikuma's name. They leased about 10 acres of land, "including the one in the mountains, a few on the top." He says, "Though I worked for ten years, I lost money each year for ten years."

Everyone had a hard time. Most people could not afford to pay the full rent in advance. Usually they paid half at the beginning of the season and paid the rest after harvest. Masa Kobayashi explains how they paid:

It [the rent] was a set price. At the end of every year, in December, we had to pay the rent fee. We went to the bank to borrow the money for it.

She continues:

Finally, the Shikumas told us it was better to move out of that place. We answered, "No, we will not move out." We held on to that land and stayed there for ten years until the war started.

Mack Shikuma, unlike his father who was active in the community, worked in the fields. Sharecroppers "just survived," he says. "Owner takes half, and half goes to half-share farmers who had to hire pickers. They paid pickers, and nothing was left."

Eiko Tsuyuki and her husband became sharecroppers after she had a baby. They managed 2 acres of strawberries and 25 acres of tomatoes. She used to exchange labor with the other sharecroppers.

We hired Pilipinos for tomato picking, and I did the packing. While I worked for the white family, the truck came at night and sent them to San Francisco. We hired Pilipinos for farm labor, but there was always something to do. I used to do the sorting and helped pick the neighbor's strawberries.

They relied upon the bosses for shipping and lost most of any profit in the process. Eiko Tsuyuki was once a sharecropper for the Shikumas.

The Shikuma family managed the marketing and transport of strawberries to San Francisco by train. For tomatoes, the landowner managed the marketing, gathering all the crops from us. They marketed them and took the profit. They just showed us what they did every day.

The local farmers relied upon the Issei for productive crops and denied them the opportunity to escape their status as sharecroppers by controlling the marketing and transportation of the goods.

The sharecropper's life was not easy. Toshi Murata was still a young man without responsibility and could enjoy traveling with the money he made one season until his money ran out. "People who were growing strawberries could not live well.... My income was higher than theirs."

Though he went many places, he "did not have much personal contact with white people." Pilipinos and Mexicans worked in the fields, and he often worked in the packing sheds. "Most of the time, Japanese stayed in the same place together." He met many single men, some old, and others young like himself.

Well, if I think of an old man, there were many Issei, not Yobiyose, pioneer Issei who came to the U.S. first.... After all, because of the wandering life, I mean, Shanghai Bank.

Instead of saving money, these Issei men lived for the day and lost, gambling away their lives. Time passed very quickly and before they knew it, they could no longer continue. They found themselves growing old or sick and began using the phrase "fujimigoro" which literally meant "this is the time to see Mount Fuji in Japan." It was used for an old man who could no longer take care of himself and had decided to return to Japan to be part of the earth.

One person I knew graduated from Waseda University and was working like that.... But he went back home to Japan... he liked to drink.

Mr. Murata saw another Issei who became sick and died in the Santa Cruz County Hospital. "He drank too much and got tuberculosis. He was sent to the county hospital." Single men who did not have money or close relatives were sent to the county hospital where they were treated free of charge. After he received a call from the hospital that the man was dead, "everybody got some money together and had a funeral for him."

Through hard times, a meager but independent farmer like

Shuro Kobayashi watched his situation follow a vicious cycle of despair.

> There were some social reasons that farmers could never get on in the world. They were always poor. They could not help but borrow money from the company at the end of the year, and so it goes on and on.

Growing Up in Watsonville

> There was blood on the face of the Nippon sky this morning when Mrs. Moromoto, a berry grower, engaged N. Iwanaga, a competitor in the business, in a battle royal on south Main Street.
>
> Mrs. Moromoto took Iwanaga to task for selling his product to the local dealers at a cheaper price than she was willing to take and which caused the market to slump.
>
> Iwanaga told Mrs. Moromoto that the berries belonged to him, and that he could sell them for any price that he saw fit, whereupon Mrs. Moromoto showed her pugilistic ability by swinging on Iwanaga's jaw with several well-delivered jolts that brought the stars of the firmament out in well developed brilliancy.
>
> Iwanaga retaliated by landing several times on Mrs. Moromoto's charming face (*Watsonville Evening Pajaronian*, June 15, 1923).

This kind of battle was probably a rare incident, especially between a man and a woman. But women were as important in business as men, and they did not always maintain a modest, gentle femininity. Mack Shikuma, who grew up in the berry fields in Pajaro, simply said, "credit it to Mother." While his father was occupied with community events, running around almost every night helping other people, his mother worked.

> Father liked to take time to do business for others. For example, for the Japanese Association, churches. Issei had family trouble, then he goes out to straighten that out.

It was a matter of course that both parents in the family worked. But a large portion of the men's work was modernized; tilling, leveling the land, and transportation were done by modern machines after 1920. In 1917, the legislature passed a fresh fruit and vegetable standardization law which enforced the quality of berries in containers. The "face pack" method (all the strawberries on top of the container facing the same direction to look attractive) was prohibited. Various new containers were tried. Wooden chests which had to be returned to the grower were gradually no longer used, as it was too costly to return them. Heavy physical labor, such as piling 150-pound chests on a wagon, also disappeared. Light dispensable crates without chests came into use. Berries, formerly transported by freight after being carried to the station by wagon, were now transported by truck.

> In recent days, there were many who used the truck. Local white berry growers like Mr. Lighter, Hopkins and Eaton were planning to transport [berries] by trucks (*Shinsekai*, Jan. 11, 1919).

Later, many adult Nisei became truck drivers in the 1930s.

On the other hand, much of women's work remained the same. Weeding and picking was done by hand in a squatting position as before. Chizu Matsuoka explains:

> The work was not heavy for women, though they had to endure a terrible backache for two to three days or a week when they started picking berries. Picking was difficult work because they had to stoop constantly.

While mothers worked all the time, they also had to raise their children one after another.

When Chizu Matsuoka started her midwifery business in 1925, the Issei were already registering their children's births at the county office, and the number of women who received assistance from a doctor or a midwife at the time of delivery increased. Although there was another Japanese midwife, Osuji Enomoto, Mrs. Matsuoka was the only midwife certified to report births to the county.

> I had a connection with the hygiene department [Department of Public Health]. I handled the birth certificates. I filled out the form for the birth certificates, brought it to the hygiene department to get permission, and sent them to Santa Cruz County.

Many people, especially in the country, still had deliveries at home and waited for a doctor or midwife to visit them. Chizu Matsuoka opened a maternity hospital for Issei women. She explains how she handled her hospital:

> When I operated my business as a midwife, I hung our signboard in

the alley behind the Buddhist church. After I moved to the present house, I opened the maternity hospital.

In the 1930s, some women had their babies delivered in her hospital, and she was the busiest before World War II, when the Yobiyose had their children. Her hospital was just a regular house with several bedrooms. She says:

> My customers were mostly Japanese, some Pilipinos, and a few Mexicans who married Japanese just before World War II. Many women visited me during the pregnancy, though I had to visit some people myself. I had to keep an eye on the baby's position. I tried to stay around home and wait for the call when I knew somebody was going to have a baby. Many women came to my place to have their babies. Sometimes, I had four or five women at the same time; we had to sleep in the living room. I charged $2.50 for all day. It cost half of what doctors charged at that time. Women usually stayed at my place from five to six days or maybe less than a week. It was nice for them to stay at my place because I served them Japanese food.

In the country, children played in the fields where their parents worked. Both at home and in the field, the people around them spoke Japanese, and they learned Japanese as their first language. The majority of the Issei were tenant farmers and made their own colonies under the same boss.

Ichiro Yamaguchi grew up in a small colony where his father leased about 20 acres of land from a local white farmer and employed four Issei sharecroppers. This was the world he grew up in.

> You know, we were brought up among Japanese. If you stay in camp, nobody speaks English. Everybody talks to you in Japanese, you had to answer in Japanese.

He was the oldest Nisei in that colony, and when he started elementary school, he had a difficult time. "I was the only one [Japanese American] in that school. Oh, it took only about two or three years [to communicate with other students]." Being casual and frank, it seems he did not have a difficult time mingling with the other students. But school work was not easy for him. "I never did catch up. Probably behind two years all the time. I was not a good student. I never liked to study." His mother became sick after delivering another child when he was about 4 years old, and at an early age he had to help his father before and after school. He also did a lot of cooking. He and his younger brother contracted pneumonia from working too hard.

Frank Sakata, who moved to Watsonville during his first year of grammar school, also had a hard time. But he says he was lucky.

> At first, I had a hard time speaking English; I could converse by the time I was in the second grade. I was lucky to have one teacher who let us stay after school to practice pronunciation.

School in the country was a one-room building where one teacher taught. At first, the one-room school Ichiro Yamaguchi attended had thirty children and one teacher. Two years later, a new strawberry field in the same district brought a number of Issei families with their children. Mr. Yamaguchi says:

> They brought in another teacher. See, that was a big problem in the school district because Japanese worked on the strawberry farm for two, three years, and they all moved out again, and there was just an extra room left there they had to do something with.

Opening of Roach School
from Murakami collection, 1931
reproduced by Media Services/U.C.S.C.

Many children had to move from one school to another, following their parents. Masao Kimoto's children also had to face moving.

> At that time, I was a hair's breadth away from bankruptcy. I could no longer be an independent farmer and went to work for someone else. I worked for the company for three years. This forced my children to change their school twice. Whenever they moved to a new school, they soon showed that they were good students as they studied hard. The teachers said that my children did so well in school that they could not teach them in their classes. They requested us to promote them one year up... my older daughter and my boy were all one year ahead from the other children.

He was lucky to have children who could manage to keep up their studies while changing schools. In reality, many children had to drop back a grade, as they could not catch up with the school work.

When Oak Grove Berry Farm had a number of Issei sharecroppers on their strawberry farm, the nearby school district (in Natividad in Salinas Valley) was occupied predominantly by Japanese American children.

> Out of 51 children attending the Natividad school, 43 are Japanese and 8 white. The school district territory comprises the Oak Grove berry ranch, said to be the largest single berry raising farm in the world, which is operated exclusively by Japanese and their families (*Watsonville Evening Pajaronian*, March 14, 1923).

Similar situations were found in the Watsonville area. Mack Shikuma, whose father hired many sharecroppers on his farm, went to school with a number of Japanese American children. He did not need to speak English in the grammar school; therefore, he had a difficult time in high school where he had to speak English. Frank Sakata remembers:

> Of 100 graduating classmates in high school in Watsonville, there were only five of us who graduated together. Two were Americans, and three were Japanese. A lot of them dropped out or moved away.

Growing up in town was different from growing up in the country. In town, the children had more chances to meet with people outside of the Japanese community. Especially after 1924, the Japanese residential area of Watsonville became a multi-ethnic town: White, Chinese, Japanese, and later Pilipinos. Those who grew up in town were more likely to be comfortable speaking English than their counterparts who lived in the countryside. Yuki Torigoe talks about her son:

> It must have been very hard for him to learn English. When my big boy started to go to school, he once told his papa, "Papa's English is different from teacher's English." [Laughter.] Papa said, "Then you should learn it by yourself." After that, he learned it by himself all right.

She tells another story about her son:

> He was very happy to go to school. Once he asked me to make a sandwich; so I made it for him. But he said that the one given by his friend was different from his. Next day, he got a sandwich from his friend to bring it back home. He asked me to make a sandwich exactly like this. [Laughter.]

The *Watsonville Evening Pajaronian*, the local English newspaper, did not neglect the Kasei Japanese American baseball team in its sports column in the 1930s. They were a strong team and practiced at Kasei field along the Pajaro River between Union and Main Street.

Ichiro Yamaguchi remembers when he played the game:

> After I quit, two or three, four years later, they became good players.... Every time we hit a ball too far, you got lost in the willows. [Laughter.] We could not find a ball. Well, we had George Shikuma, Charlie Shikuma's brother. He was one of the pitchers. He was really good.

Baseball gave country children an opportunity to meet town children. Ichiro Yamaguchi says:

> I got to know a lot of them through games because they were town people anyway, like Joe Morimoto, Charlie Iwami. They were all baseball players.

In the beginning, young Issei as well as elementary school children played the game. During the heyday of the Kasei baseball team, the members were mostly high school students. The Kasei Bees, a basketball team, was also mentioned in the paper.

The childbirth rate increased after 1912; subsequently, the number of Japanese American children increased. At first, older

Nisei had language problems in school, but soon, the honor roll in the *Watsonville Evening Pajaronian* listed many Japanese American names. The *Pajaronian* had a special column for announcements by the Japanese community. To some extent, the Japanese Americans were accepted as Americans, as part of the whole community in Watsonville.

Kibei Nisei

Many Issei had difficulty deciding whether to return to Japan. For some, time passed them by before they could decide, and their minds swung right and left, influenced by their experiences in this country.

A large number of Issei definitely wanted to return home, especially in the beginning, and they were concerned about their children's education. *Nichibei Shimbun* (May 2, 1913) commented:

> There were some people who wanted to give their children supplementary Japanese language lessons along with the attendance to white public schools, and other people who wanted them to have pure Japanese style education, as they were planning to go back to Japan sometime in the future. It could not be helped to have two split expectations on education since some parents were determined to stay, and the others did not.

Yayoi Japanese Kindergarten was established in 1912. They celebrated the first anniversary on April 28, 1913, with their parents, 25 women and several men (*Nichibei Shimbun*, April 29, 1913). The kindergarten teacher, Katsuko Asaga, the wife of the Rev. Hikoshiro Asaga, continued her work for over ten years. The Buddhist church also established the Japanese language school in 1912, which was later managed by the Japanese Association.

Other parents began to send their children to Japan to be educated, believing that their future lay in Japan. Yuki Torigoe's cheerful, lively voice becomes soft and low when she talks about it.

> By the time of the Second World War, I was still of some mind to return to Japan, I think. I wanted to go back home. Because of that, I sent two of my children [two daughters out of her four children] back to Japan to be educated in Japan, because we planned to go back someday. Everybody thought so at that time. It was wrong to live sepa-

rately from our children, though. I took my children with me to Japan, and I left them there.

Her family had strong ties to Japan. When she returned there in 1923 with her two children, they rebuilt their house in the country. She remained in Japan for two years and then returned to the U.S. Now her second daughter lives in that house in Japan.

Masao Kimoto, who was determined to stay in the U.S., explains:

> There may be some people who did not come back [from the trip to Japan] if their family at home was rich. In reality, less than one of ten immigrants came from a rich family. I have no hesitation to say that. You know, it should be clear that they would not come here if they were rich back there. They came because they were poor.

His wife, Yoshino Kimoto, explains her decision about their children:

> Before the war, there were many people who thought of going back to Japan. We were here from the time of our parents' generation, and we cleaned up everything in Japan to move here. We did not have any intention of going back. That's why we bought our own house earlier. But once I thought of returning in the future and taking back my children to Japan; so they could have a Japanese education. The relationship between parents and children became difficult when the parents stayed here in the U.S., leaving the children in Japan. Though I took them back to Japan, I came back with them again. When I sent the older one to Japan and left the younger ones at home with me, my love naturally went to the children I was with every day. That causes a gap between the parents and the older children and among children. As soon as I realized that was not good, I went to Japan to bring back the one I left in Japan.

Even those who decided to settle here from the beginning occasionally wondered if they should go back to Japan.

The Kibei Nisei are the Issei's children who were sent to Japan to be educated. They kept their dual citizenship; that is, while they registered their births in the U.S., they also kept their names in the Japanese census registration. They stood on the borderline between two countries, just as their parents' decision swung between the two countries. But because of their Japanese education, many Kibei Nisei retained Japanese cultural values.

Yoshito Kadotani is a Yobiyose who is gifted in both languages. He was brought to the U.S. at the age of four and was educated in both Japan and the U.S. He says:

> My father thought that we needed to educate children in Japanese as we are Japanese. He took the whole family back despite the sacrifice he had to make. Because of this, I am thankful to my father.

Kibei Nisei generally have a different opinion about their education in Japan. Kimiyo Kadotani, Yoshito's wife, says:

> I went back when I was 5 years old. I was not in kindergarten yet, just nursery school. I was brought back to Japan by my parents. Though my parents came back here, I did not get any American education. I was only educated in Japan. Things did not go well when I returned to the United States.

Though they were U.S. citizens by birth, they had a hard time living in the United States.

World War II

The world economy was hit by the Great Depression in 1929 and did not recover until World War II started and defense production increased. In the 1930s, while anti-Japanese feelings had abated somewhat in California, the Japanese army's activities were in the world's eyes. Antagonism and prejudice against the Japanese never ceased.

The 1930s was a time for the coming generation, the U.S. citizens of Japanese descent. By 1940 in California, Japanese Americans born in the U.S. numbered 60,148 and foreign-born numbered 33,569. In Santa Cruz County in 1940, 931 Japanese Americans were born in the U.S. and 370 were foreign-born. Many pioneer Issei were getting close to retirement age. In 1941, Kumajiro Murakami was 60 years old; Bunkichi Torigoe was 59; Unosuke Shikuma, 57; Harry K. Sakata, 56; and Shuro Kobayashi, 54.

Even though they had easily spent 40 years of their lives in the U.S., their hard work often did not bear fruit, and many were not yet completely independent farmers. In *Agricultural History*, Masakazu Iwata (op. cit., p. 32) examined the situation of Japanese farmers in California:

In 1910, eighty-five percent of them were classified as tenants and thirty years later, in 1940, seventy percent of them were still in the same category.

Adon Poli also notes in her book, *Japanese Farm Holdings on the Pacific Coast* (1944, p. 2):

In 1940, 70 percent of the Japanese farmers in the three states [California, Oregon, and Washington] were tenants, as compared with only 19 percent of all farmers.

In the 1930s, Japanese American farmers cultivated a variety of crops, especially vegetables. Census reports indicated that lettuce production jumped from 16 acres in 1920 to 2,683 in 1930 and to 4,840 in 1940 in Santa Cruz County.

According to R. Albaugh in *California Cultivator* (Jan. 18, 1930, and Aug. 2, 1930), lettuce cultivation started in Pajaro Valley in 1916, and in the 1930s it became one of the major crops in the Salinas-Watsonville areas. In 1940, Monterey County, where the Salinas Valley is located, produced the most acres of lettuce in California, 44,001; Imperial County produced 16,380 acres; and Santa Cruz County produced 4,840 acres. Watsonville was integrated with Salinas lettuce production, and many Issei in Pajaro Valley were involved. Harry K. Sakata had been one of the big Japanese lettuce farmers in the Valley, and he became a partner of the Travers Brothers in the packing business.

Travers and Sakata have taken over the second largest lettuce packing plant in Watsonville, S.C. County, and will ship ten cars a day (*California Cultivator*, Feb. 29, 1936).

Nevertheless, as Whitney indicated in *California Cultivator* (April 4, 1932), "Berries have long been a major crop in the county." In the 1930s, Issei sharecroppers were still growing strawberries for independent farmers; they had not yet escaped their dependent situation. In 1940, Los Angeles County produced 1,120 acres of strawberries, Orange County produced 515 acres, and Santa Cruz County, 422. When the Issei and the Nisei of the Pajaro Valley were forced to move to a concentration camp in Arizona during World War II, lettuce production dropped from 4,840 acres in 1940 to 2,741 in 1945, and the number of farms dropped from 75 to 71 in 1945. Strawberry production experienced even more drastic changes. The number of farms dropped from 71 in 1940 to 10 in 1945 and from 422 acres in 1940 to 8 acres in 1945, according to the census in Santa Cruz County.

The way to independence was not easy. As only a few Nisei had reached legal age, Mack Shikuma and Ichiro Yamaguchi let people use their names to lease contracts for farmland many times. Ichiro Yamaguchi laughs at how often people used his name.

> Well, I was not old enough for the last time we leased the land, we used Mack Shikuma's name. After that when I became of age, I had a lot of property under my name [which I had] nothing to do with.

Like Shikuma, he received no profit or benefit from doing this. He said, "I just do a favor. They asked, so I signed it. That's all."

By 1940, a large number of Nisei had reached legal age. Mack Shikuma was 38 years old, married, and living with his parents in Watsonville. Pooling their money, he and his father bought 40 acres of land in 1941. Few Issei were completely independent farmers before World War II. Approximately ten Issei owned their land right before World War II began. Most Issei were still in the middle of their struggle, while some had almost reached their goal.

Ichiro Yamaguchi had been a hired worker and became a sharecropper at the age of 17 when his father fell ill. It was difficult for him to become an independent, leased farmer.

> Dad was a boss, but he got sick. We had no choice; we went into sharecropping. Just until the war, we could not get out of that. I was going to get my own after I got married.

Finally, he managed to lease 20 acres of land just before the war started.

Shuro Kobayashi, who was never successful but survived to be an independent leased farmer, was happy when his son, Tetsuya, finally became 21 years old in 1940. His wife tells me:

> When our child finally graduated from high school and reached mature age, we told each other that we could now focus on the farm. Then the war started. It was a very severe thing for us. I really think that if the war had never happened....

Masao Kimoto, who became a hired worker, slowly accumulated capital to become independent.

> I was working for others after quitting my own business. I was prepar-

> ing for my own business, buying a tractor and truck, one after another. I had about $3,000 worth of them with me at that time.... It was $3,000. At that time, people got paid only 35 cents an hour. It took me a long time to get $3,000.

Then the war came. He had to sell his machines for about half price before he left for the concentration camp in Poston, Arizona.

Toshi Murata, a traveler, finally decided to settle down in Salinas five or six years before the war began. He packed lettuce for a Japanese company in Salinas. For many years, he could not find a suitable marriage partner. After the Immigration Exclusion Act in 1924, it was impossible to find a Japanese woman of his age in California. "I thought about it [marriage] many times, but I did not have any chance." Finally, he married a young Nisei woman named Kazue, arranged by a go-between in 1941.

Such were the times. Pearl Harbor was bombed on December 7, 1941. The Japanese Americans' surprise, disappointment and frustration were enormous. Suddenly, the Issei were enemy aliens. The Issei greatly feared for their situation. They were alone without protection in an enemy country.

Pilipinos were at the bottom of the social ladder in California agriculture, as the Issei were formerly, and many were hired by Japanese American sharecroppers or leased farmers. After Japan attacked the Philippines, Pilipinos were antagonistic toward the Japanese Americans. Eiko Tsuyuki, a sharecropper, says:

> They [Pilipinos] threw stones at the windows. Japanese girls who were about that age, going to high school, were bothered by them. They really did terrible things. They even came to the country. Fortunately, none of our family was bothered by them.

The attack on Pearl Harbor was a great surprise to many people, but what followed was even more difficult to bear. The day after Pearl Harbor, many community leaders were taken away by the FBI. When Bunkichi Torigoe, who sold guns at his store, was taken away, his wife, Yuki, temporarily lost her voice. Soon all the Issei's money in the banks was frozen.

> Funds of all Japanese in the United States were frozen in an executive order from the treasury department. Pending receipt of further instructions, no funds may be withdrawn from banking institutions (*Watsonville Register Pajaronian*, Dec. 9, 1941).

Restriction on movement was enforced, and the Issei were forced to move inland from the coastal area roughly bordered by State Highway No. 1 (*Watsonville Register Pajaronian*, Feb. 4, 1942). Motoki, the secretary of the Japanese Association, responded in the same article:

> "We must comply with the law," said Motoki, "but we don't as yet know how." He pointed out that there is a serious shortage of houses available to Japanese on the east-of-Main street side. Twenty-three families within the city limits would be affected by the order.

Masaki Yamaguchi was one of them. His son explains how they managed to move:

> Yeah, we were on the coast side, see. Do you know where the Tano's house is? On Freedom Boulevard where you take it to go to Santa Cruz, where you go to Corralitos? From that junction, we were on this side, way back, below the foot of that hill. You go in between the service station and garage. He [his father] had to move into the house we rented. A whole bunch of Issei got together until we had to move out [to camp]. So all Nisei who were living over on this side, we had to pack up. We had to do all the packing by ourselves.

A number of families had to share a single space. Few local people were willing to rent a place to enemy aliens. This evacuation was allowed from Feb. 17 to Feb. 24, 1942.

Final evacuation for all Japanese Americans, born in the U.S. and born in Japan, was announced. Everything had to be prepared by April 30, 1942 (*Watsonville Register Pajaronian*, April 21, 1942). They were given only ten days to sell or store their belongings. Large and small items were sold for a mere fraction of their value. They were allowed to take only what they could carry to the concentration camps.

After the fleeting days and nights in confusion, they thought of one thing: all their work and dreams for many years were reduced to nothing. As the *Watsonville Register Pajaronian* announced on April 29, 1942:

"Japs Disappear From Valley As Evacuation [Is] On."

Looking Back

After World War II, some Issei returned slowly to Watsonville from the concentration camps. Some found their belongings stored at churches and at their homes stolen or lost, while others had their belongings returned by trusted neighbors. In general, most Issei who were sharecroppers went elsewhere, but those who had left some property came back. The Asian American section of town was never re-established after the war. Japanese Americans started their lives all over again.

After 1952, Shin Issei moved into the Pajaro Valley as agriculture continued to be successful. By the time I started this project, the older Nisei, some in their 60s and 70s, were retiring or continued to manage farms, some of them prosperous ones. These Issei whom I interviewed were between 70 to close to 100 years old and lived by themselves or with their children.

The Issei now face old age, and the language barrier creates even more isolation. One Issei woman told me, "I could not do much, but my children made progress." At the same time, she said something else; it was so short and insignificantly spoken that it slipped my mind. After a few seconds, I was struck with the weight of what she said: "I am lonely."

Issei in Watsonville could marry and establish themselves in this country as the forebearers of Nisei, Sansei, and following generations of Japanese Americans. But their daily association has been quite isolated from other races; the majority of Issei dealt with the local white population only through business. Unosuke Shikuma was one of the few Issei who had close contact with the local white people. While his son, Mack Shikuma, said, "they [his parents] always intended to go back to Japan," Hiroko Shikuma, his wife, remembered that her father-in-law "always said that this is my adopted country, so that he was going to be faithful to it."

Ichiro Yamaguchi, an older Nisei, said, "he [his father] is still Japanese. I am still Japanese." When I said that he is Japanese American, he answered, "I am American, but still Japanese," and continued:

> See, up until I started gardening, I was more Japanese. When I started gardening, I got to understand Caucasian things and began to realize they're not much different from my own, you know. Up until I started gardening, I did not know what they are thinking. Now I get to make contact with all those white people, I know they are not thinking much different than you and anybody else.

There was one main reason why Masaki Yamaguchi never thought of going back to the country where he was born. His son Ichiro explained it:

> He had no interest in Japan any more. He did not even go to visit. I asked him once, since he was the only person I knew who never visited Japan since he left.... After the war, I asked him if he would like to go back. Even if he went, he does not know anybody there. Even if they are related, it would be outdated. If you see somebody on the street, "Oh yeah, you are related." "Yes." That's about it. You are not acquainted. So he said.

Masaki Yamaguchi lived to be 102 years old and passed the American citizenship test just before his death.

Shuro Kobayashi and his wife, Masa, are both Christians. Mr. Kobayashi was bitter about the United States at first.

> I did all the suffering possible here. Many of my classmates are established and became famous, but I came to the United States and stayed poor for 70 years.

His wife, Masa Kobayashi, had similar feelings in the beginning, but she said:

> After we came to Watsonville, we worked once on a small farm in

Aromas, 8 miles from here. Reverend Koga's father recommended us to that farm. We could not have our own farm, instead we worked under the white people with a share contract. That time, I met a good couple as neighbors, whose name was Mr. and Mrs. MacNair. The husband was German and the wife was English. We lived as neighbors for five years. Because they were such nice people, I began to like the United States.

Then her husband told his story:

When right after the war we came back here, we rented the house from the neighbors who were German, the owner of the house. They were German, and I am Japanese. We probably shared something in that we understood each other. His back hurt so much that he could not take care of the chickens, I mean, he could not clean up the chicken house. I was still young at that time. It was an easy thing to clean the chicken house. Without compensation, I just kept cleaning his chicken house. Before we knew it, we became friends. Once I asked him, "Could you rent this house to me?" He told me he would sell it to me. We did not have any money at all, but he did not care whether I had money or not. I, therefore, borrowed the money from the bank and bought this house at a cheap price. There were many different people in the United States!

He remembered another story:

When my children went to school, children called them "Jap." When I walked by the school, I heard them calling [my children] "Jap." I got angry at them and went to the school. Then surprisingly, the schoolteacher to whom I complained did not treat me as a mere "Jap." Instead she told me that I was right. This schoolteacher and my wife became friends. She really took care of us.

Masa Kobayashi continued the story:

I trembled to think how it would come out. The principal of the school was Mr. Lee, who was a fine man. He told us my husband was right and scolded the students. After that, a kind lady, one of the neighbors who was very nice, took me to the PTA meeting, as well as the cooking course in school. I was asked by the principal of the school to make a speech in English.

Unfortunately, she did not do it because she thought she lacked good English ability.

Masa Kobayashi and Toshi Murata were the first Issei to receive

U.S. citizenship in the Watsonville area, and many Issei followed. At the same time, they withdrew their Japanese nationalities. Toshi Murata, who had a rather clear idea from the beginning that he was going to stay here, made his final decision in 1932.

> I visited Japan as my mother wanted me to come back. In Japan, I realize that this is not a country where I can live. I was determined to stay here at the moment.

Many Issei decided to stay in the U.S. because of their children. A daughter of Kumajiro Murakami told me, "He decided to settle down in the U.S. due to the children." Eiko Tsuyuki told me of her case:

> I had never thought of going back to Japan. Once I went back to Japan with my older daughter when she was 12 years old. She did not like the country in Japan at all. She told me that she would like to go back to the United States by herself if her mommy took her to the boat and her daddy was waiting at San Francisco if mommy did not want to go back. I thought I would give her a Japanese education if she liked Japan. I gave it up. That was the chance to make my decision. I gave all my portion of the inheritance to my brother.

Looking back on it now, most of the Issei I talked to did not complain bitterly about the racism and discrimination they experienced in the U.S. When I asked one of the Issei if he was ever called "Jap," he replied, "I can't pay attention to such a little thing!" Even though they had gone through unbearable stress, many Issei would simply laugh at such incidents as if they were telling me funny stories. Their pride and confidence in themselves as survivors helped free their spirits from dwelling solely on the bitterness of their experience here. Tokushige Kizuka said of his position in this country:

> Isn't it good that we are Japanese? The United States is composed of people from various countries. Isn't it necessary for us to express the characteristics of Japanese here?

Postscript

I started this project without having any prior experience. I was born in Japan and became a permanent resident of the U.S. in 1975. As it is always difficult to adjust to a new country, I thought I would remain a "foreigner" and would be unable to find my own place. However, I got involved in this project. Since then, there has always been something that has made me continue.

It was a great challenge to interview someone for the first time. I found that it was a heart-warming experience. A very personal communication was evoked not just by the words that were spoken, but through continued correspondence, as well as reading notes and listening again and again to the taped interviews. It humbled me and made me thankful that people older than my parents trusted me and told me of their lives in this country.

It seems a long time ago that I started this project in 1978. Many of the Issei I interviewed have weakened in physical strength. Mr. Shuro Kobayashi passed away on June 2, 1979. He was 92 years and 7 months old. Life in the United States has been long for Kumajiro Murakami, who in 1978 was 97 years old. He walks an inch at a time, slowly but surely. He can still hear well, but he is losing his eyesight. When I told him that I was leaving for Seattle, he looked sad and, with a red face, shook my hand with his two hands. On the way from his house to the bus stop, looking up at the clouds starting to color the evening, I thought all the houses around me looked vivid and bright.

During a sunny day in January on one of my later visits to Watsonville, I happened to see Mr. Murakami closing the back gate of his house. I started to run and called to him, "Kumajiro-san." He moved slowly, but he recognized me right away and smiled. "Welcome, welcome. Come on in, come on in. Let's make some time to talk, let's make some time to talk!" When I was going to follow him, I saw his wife, Fuji Murakami, working in the garden. I knew she was sensitive and that she tended to feel left out of conversations as her hearing was getting very bad. She was 93 years old at that time. I saw her walking quite fast with big strides. Oh my, how high she could lift her leg when she kicked a big, fat cat in the stomach who ran in front of her. Her back was slightly bent, but she could still take care of their household. It had started to rain, but I waited outside until she came to the house. Upon her approach, she looked at me with a little approval and welcomed me inside.

The next year when I came back, I heard a hoe softly breaking dried earth behind the bushes near their house. I recognized Mrs. Murakami. I went around and stood where she could see me. She said, "Oh, welcome back, Mrs. Nakane." Later when I was watching television with them, she said, "I would like to die as soon as possible because my back aches." Another time she said, "I feel sorry for Papa if I go because I would not like to put Papa in the nursing home."

> Papa always wakes up in the middle of the night and takes a nip of *sake* and leaves it in the kitchen. I can tell. When he has a scary dream, he comes to wake me up.

She passed away on Aug. 21, 1983. She was 95 years old.

Once I brought a friend to take a photograph of Mr. and Mrs. Murakami. He dressed up especially for the occasion in his rather dark suit, and she looked somehow smaller, scrunched up in formal clothes. In front of their living room window, they patiently followed my friend's requests to move slightly right and left and waited for the clicks of the shutter. When it was finally over, Mr. Murakami leaned on the sofa decorated with her knitted laces, relaxed both his arms, and his face became alive. He said to me, "I will sing a song for you." He could sing well; only the volume of his voice was weakened by his age. After his singing, I said to him, "Please live long." "Don't worry," he said, "I will not die." And he laughed.

Kazuko Nakane

Fuji and Kumajiro Murakami
photo by Mona Nagai, 1983

Interviewees

Yoshito Kadotani with Kimiyo Kadotani
Shizuko Kajihara
Masao Kimoto with Yoshino Kimoto
Tokushige Kizuka
Shuro Kobayashi with Masa Kobayashi
Miteru Mano
Chizu Matsuoka
Kumajiro Murakami with Fuji Murakami
Toshi Murata
Kokuhō Nakamura
Frank Sakata
Mack Shikuma with Hiroko Shikuma
Yuki Torigoe
Eiko Tsuyuki
Ichiro Yamaguchi
Kenzo Yoshida

Footnotes

1. Y. W. Abiko of *Nichibei Times,* who was the only son of Kyutaro Abiko, told me that only a limited number of Japanese names are found in this book because some Issei did not bother to be listed, as payment was involved.
2. Because of limited information, the Military Archives Division of the National Archives and Records Service in Washington, D.C., could not confirm that he served in the U.S. Navy.
3. *Zaibei Nihonjinshi* [History of Japanese in the United States]. San Francisco: Zaibei Nihonjinkai Japanese Association, 1940, p. 153.
4. Quoted in Kashiwamura, op. cit., p. 294.
5. One of my Chinese informants told me that Japanese women who used to work for the prostitution houses married Chinese men in the very early days.
6. Kō Murai, *Zaibei Nihonjin Sangyō Sōran* [A Compendium of Japanese Agriculture in the U.S.]. Los Angeles: Beikoku Sangyō Nippōsha, 1940, p. 354. They documented one Japanese man from Okayama Prefecture who raised strawberries lasted 14 years on the same land in Lindsey, in Central California.

Bibliography

Agricultural Crop Report. Watsonville: Agricultural Commissioner, 1980.

Albaugh, R. "Prosperous Monterey County." *California Cultivator* (Jan. 18, 1930): 59.

—. "The Lettuce Capital of the World." *California Cultivator* (Aug. 2, 1930): 99.

Atkinson, Fred W. *100 Years in the Pajaro Valley from 1979 to 1868.* Watsonville: Register and Pajaronian Press, 1934.

Beattie, W. R. "Lettuce Growing." U.S. Department of Agriculture. *Farmers Bulletin* 1609 (1929).

Bliven, Bruce. "The Japanese Problem." *Nation* (Feb. 2, 1921):171-172.

Buddhist Churches of America 75 Years History 1899-1974, vol. 1. Chicago: Nobart, Inc., 1974.

Build a Greater Sangha: Sixtieth Anniversary 1906-1966. Watsonville: Watsonville Buddhist Church, Oct. 29, 1966.

Bunje, Emil T. H. *The Story of Japanese Farming in California*. Berkeley: U.S. Works Progress Administrataion Project, 1957.

California Commission on Immigration and Housing. *Advisory Pamphlet on Camp Sanitation and Housing*. Rev. San Francisco: California State Publishing office, 1919.

California Horticulture Commission. "Berries as an Intercrop in Young Orchards." *Monthly Bulletin* 4 (1915):64-68.

"California Hustling Japanese." *The Literary Digest* (July 12, 1913):67-72.

California State Agricultural Society. *Transactions*. Sacramento: California State Publishing Office, 1868.

—. *Report*. 1906, pp. 219-223.

California State Board of Agriculture. *Annual Report 59*. Sacramento: State Publishing Office, 1913.

—. *Annual Report for the Year 1917*. Sacramento: State Publishing Office, 1918.

—. *Statistical Reports*. Sacramento: State Publishing Office, 1920.

Chinn, Thomas, ed. *A History of the Chinese in California: A Syllabus*. San Francisco: The Chinese Historical Society of America, 1969.

Chiu, Ping. *Chinese Labor in California 1850-1880*. Ann Arbor: Edwards Brothers, Inc., 1963.

Cikuth, P. Luke. "The Pajaro Valley Apple Industry 1890-1930." An interview conducted by Elizabeth Spedding Calciano. Santa Cruz: University of California Library, 1967.

Cox, M. L. "Watsonville's Spring Lettuce." *Pacific Rural Press* (March 22, 1924):213.

Eaton, O. O. "Strawberries in the Pajaro Valley." *California Cultivator* 58 (1922):315-318.

An Economic Survey of the Cities of Santa Cruz, Watsonville, and Contiguous Areas. Santa Cruz: Chamber of Commerce, 1944.

Erdman, H. E. "The Development and Significance of California Cooperatives, 1900-1915." *Agricultural History* 32 (1958):179-184.

Farm Editorials. *California Cultivator* (Jan. 17, 1920):78.

Fisher, Lloyd H. *The Harvest Labor Market in California*. Cambridge: Harvard University Press, 1953.

Fujita, Michinari. "The Japanese Associations in America." *Sociology and Social Research* (Sep.-Aug. 1928-1929):211-228.

Hanihara, Masanao. "The Japanese Question in the United States – The Facts of the Case." *Economic World* (March 23, 1918): 400-402.

Harrison, E. S. *History of Santa Cruz County, California*. San Francisco: Pacific Press Publishing Co., 1892.

Hill, R. G. "Preparing Strawberries for Market." U.S. Department of Agriculture. *Farmers Bulletin* 1560 (1928).

Hutchinson, Paul. "Kagawa: Proletarian Saint." *Atlantic Monthly* 157 (1936): 594-600.

Ichihashi, Yamato. *Japanese Immigration: Its Status in California*. San Francisco: Marshall Press, 1915.

–. *Japanese in the United States*. 1932. Reprint. New York: Arno and The New York Times, 1969.

Ichioka, Yuji. "America Nadeshiko: Japanese Immigrant Women in the United States, 1900-1924." *Pacific Historical Review* 49 (1980): 339-357.

–, comp. *A Buried Past*. Berkeley, Los Angeles, London: University of California Press, 1974.

Ito, Kazuo. *Issei: A History of Japanese Immigrants in North America*. Translated by Shin'ichiro Nakamura and Jean S. Gerard. Seattle: Executive Committee for the Publication of *Issei: A History of Japanese Immigrants in North America*, 1973.

Iwata, Masakazu. "The Japanese Immigrants in California Agriculture." *Agricultural History* 36 (Oct. 1962): 25-37.

The Japanese Farmers in California. San Francisco: Japanese Agricultural Association, 1918.

Johnson, Arthur T. *An Englishman's Impressions of the Golden State*. London: Microfiches, 1913.

Johnson, Eleanor. "The Japanese Americans in the Pajaro Valley – First Immigrants, Organizations, Evacuation, and Contributions." Unpublished paper. Santa Cruz: University of California Library, Sept. 1, 1967.

Kaneto, Shindo. *Matsuri No Koe* [The Sound of Festivals]. Tokyo: Iwanami Shinsho, 1977.

Kashiwamura, Kazusuke. *Hokubei tōsa taikan* [A Broad Survey of North America]. vol. 1. Tokyo: Ryūbundō, 1911.

Katō, Bungo. *Saikin No Zaibei Dōhō* [Our Brothers in America]. Tokyo: Nihon Zusho Shuppan, 1921.

Kato, Shin'ichi. *Beikoku Nikkeijin Hyakunenshi* [One Hundred Years of Japanese American History]. San Francisco: Shin Nichibei Shimbunsha, 1961.

Kawakami, K. K. "California and The Japanese." *Nation* (Feb. 2, 1921): 173-174.

Kawamura, Yusen, ed. *Beikoku Kashū Nihongo Gakuen Enkakushi* [A History of Japanese Language Schools in California]. San Francisco: Hokka Nihongo Gakuen Kyokai, 1930.

Kelsey, Carl, ed. "Present-Day Immigration with Special Reference to the Japanese." *Annals of the American Academy of Political and Social Science* 93 (1921): 1-232.

Koga, Sumio, ed. *A Centennial Legacy: History of the Japanese Christian Mission in North America 1877-1977*, vol. 1. Chicago: Nobart, Inc., 1977.

Laing, Michiyo; Laing, Carl; Takarabe, Heihachiro; Tokuno, Asako; and Umeda, Stanley, eds. *Issei Christians: Selected Interviews from the Issei Oral History Project*. Sacramento: Issei Oral History Project, Inc., for the Centennial Celebration of the Japanese Christian in North America 1877-1977, 1977.

Marlatt, Daphne. *Steveston Recollected: A Japanese-Canadian History*. Victoria: Provincial Archives of British Columbia, 1975.

Martin, Edward. *History of Santa Cruz County, California, with Biographical Sketches of the Leading Men and Women*...etc. Los Angeles: Historic Record Company, 1911.

–. "The Story of Watsonville in Its Early Days." Mimeographed from the *Watsonville Register-Pajaronian*. March 30-31, April 1-3, April 6-10, 1964. Santa Cruz: University of California Library, 1964.

Matsui, Shichiro. *Economic Aspects of the Japanese Situation in California*. Masters thesis. University of California at Davis, 1922.

McWilliams, Carey. *Factories in the Field: The Story of Migratory Farm Labor in California*. Boston: Little, Brown and Company, 1939.

Misawa, Steven, ed. *Beginnings: Japanese Americans in San Jose: Eight Oral Histories*. San Jose: San Jose Japanese American Community Senior Service, 1981.

Miyamoto, Shotaro Frank. *Social Solidarity among the Japanese in Seattle*. Seattle: University of Washington Press, 1939.

Modell, John. *The Economics and Politics of Racial Accommodation: The Japanese of Los Angeles 1900-1942*. Urbana: University of Illinois Press, 1977.

More, C. T., and Truax, H. E. "Preparation of Strawberries for Market." U.S. Department of Agriculture. *Farmers Bulletin* 979 (May 1918).

Murai, Kō. *Zaibei Nihonjin Sangyō Sōran* [A Compendium of Japanese Agriculture in the U.S.]. Los Angeles: Beikoku Sangyō Nippōsha, 1940.

Murakami, Elaine. "The California 'Problem' of Japanese American Births." Paper for History 199. University of California at Santa Cruz, Fall 1974.

Murakami, Elaine. "California's Alien Land Law." Paper. University of California at Santa Cruz, Fall 1975.

—. "Agricultural Cooperative Systems of the Japanese in California." Paper. University of California at Santa Cruz, Winter 1976.

Naka, Kaizo. *Social and Economic Conditions among Japanese Farmers in California.* Masters thesis. University of California, at Berkeley, 1913. Reprint. R. & E. Research Associates, 1974.

Newcomb, W. I. "The Berry Industry." *Pacific Rural Press* (Jan. 8, 1910): 24.

Nichibei Shimbunsha. *Nichibei Nenkan* [Japanese American Year Book], nos. 6-8, 10, 12. San Francisco: Nichibei Shimbunsha, 1910-1912, 1914, 1918.

—. *Beikoku Nikkeijin Hyakunenshi* [One Hundred Years of Japanese American History]. San Francisco: Nichibei Shimbunsha, 1961.

Niisato, Kan'ichi. *Imin Aiwa* [Sad Tales of Immigrants]. Tokyo: Shimpōsha, 1933.

Pacific Rural Press. San Francisco: 1900-1901.

Pajaro Valley Board of Trade. *The Apple and Strawberry Center West of the Missouri River,* 1903.

Poli, Adon. *Japanese Farm Holdings on the Pacific Coast.* Berkeley: U.S. Department of Agriculture, Bureau of Agricultural Economics, 1944.

San Francisco Chronicle. 1900-1901.

Santa Cruz County Agricultural Commissioner. *Agricultural Crop Report,* 1980.

Santa Cruz County Directory. Santa Cruz: Western Directory Co., 1920, 1922.

Santa Cruz, California Chamber of Commerce. *An Economic Survey of the Cities of Santa Cruz, Watsonville, and Contiguous Areas,* Oct. 1944.

Strong, Edward K., Jr. *Japanese in California.* Palo Alto: Stanford University Press, 1933.

Tachiki, Amy; Wong, Eddie; Odo, Franklin; and Wong, Buck, eds. *Roots: An Asian American Reader.* Los Angeles: UCLA Asian American Studies Center, 1971.

Taylor, Frank J. "The People Nobody Wants." *The Saturday Evening Post* (May 9, 1942):24-25, 64, 66-67.

Taylor, Paul S., and Vesey, Tom. "Historical Background of California Farm Labor." *Rural Sociology* 1 (1936):281-295.

Terakawa, Hōkō. *Hokubei Kaikyo Enkakushi* [History of the North American Honganji Mission]. San Francisco: Honganji Hokubei Kaikyo Honbu, 1936.

U.S. Bureau of the Census. *Census Report – Agriculture.* Washington, D.C.: Government Printing Office, 1860-1945.

U.S. Bureau of the Census. *Census Report—Population.* Washington, D.C.: Government Printing Office, 1860-1940.
U.S. Immigration Commission. *Reports*, vol. 23. Washington, D.C.: Government Printing Office, 1911a: 62-89.
—. *Reports*, vol. 24. Washington, D.C.: Government Printing Office, 1911b: 431-451.
"Watsonville Local Column." *Nichibei Shimbun* [Japanese American News]. 1913, 1915.
"Watsonville Local Column." *Shinsekai Shimbun* [New World Newspaper]. 1907, 1913, 1919-1920, 1924, 1930.
Watsonville Pajaronian. 1870-1872, 1882-1884, 1900-1901, 1905.
Watsonville Evening Pajaronian. 1919-1920, 1923-1925, 1933.
Watsonville Register-Pajaronian. 1941-1942.
Whitney, D. J. "Conditions in Santa Cruz County." *California Cultivator* 4 (April 4, 1932): 337.
Wilhelm, Stephen, and Sagen, James E. *A History of the Strawberry: From Ancient Gardens to Modern Markets.* Berkeley: University of California, Division of Agricultural Science, 1974.
Williamson, Paul S. "When Is Lettuce Most Profitable?" *California Cultivator* (Aug. 11, 1928):123, 127.
Yoneda, Karl. "One Hundred Years of Japanese Labor in the USA." In *Roots: An Asian American Reader*, edited by Amy Tachiki, et al. Los Angeles: UCLA Asian American Studies Center, 1971: 150-158.
Yoshida, Yosaburo. "Sources and Causes of Japanese Emigration." *Annals of American Academy of Political and Social Science* 34 (July-Dec. 1909): 157-167.
Zaibei Nihonjin Jinmeijiten [Japanese Who's Who in America]. San Francisco: Japanese American News, 1922.
Zaibei Nihonjinkan [A Dictionary of Japanese in the United States]. San Francisco: Shinsekai, 1922.
Zaibei Nihonjinshi [History of Japanese in the United States]. San Francisco: Zaibei Nihonjinkai [Japanese Association], 1940.

Other BayTree Books

BayTree Books, a project of Heyday Institute, gives voice to a full range of California experience and personal stories.

Allensworth, the Freedom Colony (2008)
Alice C. Royal with Mickey Ellinger and Scott Braley

Archy Lee: A California Fugitive Slave Case (2008)
Rudolph M. Lapp

Edges of Bounty: Adventures in the Edible Valley (2008)
William Emery and Scott Squire

Jazz Idiom: Blueprints, Stills and Frames (2008)
Charles L. Robinson and Al Young

Tree Barking: A Memoir (2008)
Nesta Rovina

Walking Tractor: And Other Country Tales (2008)
Bruce Patterson

Where Light Takes Its Color from the Sea: A California Notebook (2008)
James D. Houston

Ticket to Exile: A Memoir (2007)
Adam David Miller

Fast Cars and Frybread: Reports from the Rez (2007)
Gordon Johnson

The Oracles: My Filipino Grandparents in America (2006)
Pati Navalta Poblete

Since its founding in 1974, Heyday Books has occupied a unique niche in the publishing world, specializing in books that foster an understanding of the history, literature, art, environment, social issues, and culture of California and the West. We are a 501(c)(3) nonprofit organization based in Berkeley, California, serving a wide range of people and audiences.

We are grateful for the generous funding we've received for our publications and programs during the past year from foundations and more than three hundred and fifty individual donors. Major supporters include:

Anonymous; Audubon California; BayTree Fund; B.C.W. Trust III; S. D. Bechtel, Jr. Foundation; Fred & Jean Berensmeier; Joan Berman; Book Club of California; Butler Koshland Fund; California State Automobile Association; California State Coastal Conservancy; California State Library; Candelaria Fund; Columbia Foundation; Community Futures Collective; Compton Foundation, Inc.; Malcolm Cravens Foundation; Lawrence Crooks; Judith & Brad Croul; Laura Cunningham; David Elliott; Federated Indians of Graton Rancheria; Fleishhacker Foundation; Wallace Alexander Gerbode Foundation; Richard & Rhoda Goldman Fund; Marion E. Greene; Evelyn & Walter Haas, Jr. Fund; Walter & Elise Haas Fund; Charlene C. Harvey; Leanne Hinton; James Irvine Foundation; Matthew Kelleher; Marty & Pamela Krasney; Guy Lampard & Suzanne Badenhoop; LEF Foundation; Robert Levitt; Dolores Zohrab Liebmann Fund; Michael McCone; National Endowment for the Arts; National Park Service; Philanthropic Ventures Foundation; Alan Rosenus; Mrs. Paul Sampsell; Deborah Sanchez; San Francisco Foundation; William Saroyan Foundation; Melissa T. Scanlon; Seaver Institute; Contee Seely; Sandy Cold Shapero; Skirball Foundation; Stanford University; Orin Starn; Swinerton Family Fund; Thendara Foundation; Susan Swig Watkins; Tom White; Harold & Alma White Memorial Fund; and Dean Witter Foundation.

For more information about Heyday Institute, our publications and programs, please visit our website at www.heydaybooks.com.